Survey of Engineering

An Introduction to Engineering & Technology for Middle School and lower High School grades

Authors:

Alan G. Gomez, MS
University of Wisconsin & Sun Prairie Area Schools

William C. Oakes, PhD
Purdue University

Les L. Leone, PhD
Michigan State University

Editor:
John L. Gruender
Executive Editor
Great Lakes Press, Inc.

Great Lakes Press, Inc.
PO Box 550 / Wildwood, MO 63040
(800) 837-0201
custserv@glpbooks.com
www.glpbooks.com

International Standard Book Number: 978-1-881018-31-5
(1-881018-31-8)

All comments and inquiries should be addressed to:

Great Lakes Press, Inc.
c/o John Gruender, Editor
PO Box 550
Wildwood, MO 63040-0550

jg@glpbooks.com
phone (800) 837-0201
fax (636) 273-6086

www.glpbooks.com

Library of Congress Control Number (PCN): 2006928760

Printed in the USA

10 9 8 7 6 5 4 3 2 1

Important Information

This book belongs to: ______________________________

____

____

____

phone: ______________________________

Contents

Preface

One of the most significant labor shortages the United States has is technologically literate people. Every year, the U.S. government accepts more people from foreign countries on work visas to place them in technology-related fields. The continued use of the H-1B visa program during one of the tech industry's most severe downturns has heightened renewed criticism of the program. It is a hot-button issue with many U.S. engineers who fear the country is giving away its tech jobs (Bjorhus, 2003). Although we are doing more than we have in the past to give our students opportunities to become technologically literate, too often educators place students in front of computers and assume that technological literacy will follow. When colleges use surveys to find out what skills incoming freshman have, including computer skills, their skills are much lower than expected. Though teachers need to teach content with computers, educators cannot teach with just computers. Students need to be given exposure to the creative nature of engineering through design projects, hands-on laboratories, and open-ended problem solving (Sheppard & Jenison, 1996).

As budgets are cut and programs follow, parents, teachers, administrators, legislators, and taxpayers want to know that students are getting the best possible educational experiences for their dollar. These "shareholders" are an integral part of education and their roles should not be slighted. "It's bound to hit K-12 education," said Jane Hannaway of the Washington-based Urban Institute, who is researching the ailing economy's impact on schools. "We're really beginning a significant, very serious period of resource trouble" (Richard & Sack, 2003). Is it entirely subjective to determine what is a good course versus a bad one?

Parents are interested in an education for their students that has more than just graduation requirements at stake. They want to have classes and experiences that will enlighten their students to some of the options they have in society. Across the country, parents are pouring back into schools, questioning what goes on in their children's classrooms and pitching in to fill vacuums created by decreased funding (Walters, 1995). Teachers in schools are interested in having academically viable courses and programs that provide experiences for students and not courses that "keep the students busy for an hour."

Administrators are interested in keeping the school running smoothly and their teachers involved with the school's direction. During the course of a day, there may be over twenty-five meetings for administrators with parents, students, and teachers. Scheduled meetings coexist with the responsibility to deal with immediate problems that spring up at a moment's notice. Though administrators ultimately make decisions about programs, they focus on that problem when it becomes time to determine allocation and scheduling. This is a result of a system that overworks administrators and still expects them to evaluate which programs are more important than others.

Legislators are interested in the results of testing. Some of the courses and programs offered in schools do not teach the content of the test. One of the largest questions remaining in the minds of educators is if these courses are incorrectly designed (they do not directly contribute to better test scores or do not teach the exact materials that will be on the tests) or if the test is incorrectly designed (it should include more reasoning, problem solving or open-ended creative questions).

Taxpayers are looking to get the best education for their money, without wasting it on bureaucratic positions and programs that drain large amounts of resources compared to all other programs in the system. The demographic fact is that only about 25 percent of households have school-aged children—a historic low. Increasingly, therefore, taxpayers are looking at education as a financial investment that benefits other people's children (Labaree 1997, 62) and not their own.

This book and its standards-based learning activities are organized around a set of concepts, skills, and attitudes necessary to be successful problem solvers in life or a career in engineering. These hands-on courses should be created and implemented with student success in mind. It is illogical to create large sections of courses to satisfy allocation issues in schools. This type of planning is at the expense of students. Laboratories employ active learning and a smaller class size to achieve two objectives: (1) to better inform students about the nature of engineering and its specific disciplines and (2) to improve these students retention in engineering (Hoyt & Oland, NDA). The first engineering schools in the United States used the laboratory as the primary mode of instruction (Durfee, 94).

Learning to navigate the road to the solution is as important, if not more important than finding the solution itself. Education needs to teach the learning process that students need to navigate that road. Typical classrooms have students sitting in rows, five across and six deep discouraging their talking with each other. Yet educators expect them to communicate and interact with people as a result. When the students graduate from high school, whether they go into the workforce, the military, two-year colleges or four-year institutions, they will have to be able to work with others, in small groups, and in teams. In a time of budget crisis or not, courses offered to high school students should be a contributing factor to their graduation and to their lifelong success. There is an increasing need for students to be successful in grade point averages and to be creative thinkers and problem solvers.

REFERENCES

Barker, Robert. The art of brainstorming. Business Week 8/19/2002–8/26/2002;3796:168.

Bjorhus, Jennifer. Layoffs at Sun prompt inquiry over work visas. Mercury News. Jun. 23, 2002.

Daniel, Michelle. When creativity counts. Women in Business. July/August 1995;47:4;48.

Durfee, W. K. Design education gets real. Technology Review. February/March 1994;42-51.

Gomez, Alan G. Engineering, but how? The Technology Teacher. 2000;59:6:17-22.

Gomez, Alan G. Foundations of Technology. International Technology Education Association and The Center to Advance The Teaching of Technology and Science (CATTS).3/2003.

Hoyt, Marc. and Oland, Matthew. The impact of a discipline-based introduction toengineering course on improving retention.

Kardos, Geza. On writing engineering cases. Proceedings of ASEE National Conference on Engineering Case Studies. 1979.

Labaree, D. F. Public goods, private goods: the American struggle over educational goals. American Educational Research Journal. 1997;34:39-81.

Lewis, Bob. Brainstorm no-brainer. InfoWorld. 9/30/2002;24:39;48.

Moreno, Roxana and Mayer, Richard E. Gender differences in responding to open-ended problem-solving questions. Learning & Individual Differences. 1999;11:4.

Richard, Alan and Sack, Joettal. States brace for tough new year. Education Week. 1/8/2003;22:16.

Sheppard, Sheri and Jenison, Rollie. Freshman engineering design experiences: an organizational framework. The International Journal of Engineering Education. August 1996.

Verespej, Michael A. Vanishing breed. Industry Week.5/7/2001;250:7.

Walters, Laurel Shaper. Christian parents finding way back to schools. Science Monitor. 2/23/95;87:61.

Wisconsin's Model Academic Standards. (1998). Wisconsin Department of Public Instruction.
http://www.dpi.state.wi.us/standards/index.html.

Wisconsin's Governor's Work-Based Learning Board WBL-10234 (R. 10/2000) Skills standard checklist.

Chapter 1

What is Engineering?

INTRODUCTION

Engineers produce things that affect us every day. They invent, design, develop, manufacture, test, sell, and service products and services that improve people's lives.

Frequently, students early in their educational careers find it difficult to understand exactly what engineers do and where they fit best in career opportunities available to engineers.

Image Courtesy of Steve Preston and the University of Wisconsin College of Engineering

Common reasons for students to be interested in engineering include the following:

1. They like math and science.
2. A school counselor suggested it.
3. They have a relative who is an engineer.
4. They heard the field has great job opportunity.
5. They read that engineers make a lot of money.
6. They always took things apart when they were younger.
7. They like to build things.

Though these can be good reasons, they do not imply a firm understanding of engineering. What is important is that a student who is interested in engineering understands what that career requires and the options it has. We have our own strengths and talents, and finding places to use those strengths and talents is the key to a rewarding career.

The purpose of this chapter is to provide information about what engineers do.

THE ENGINEER AND THE SCIENTIST

To understand what engineers do, a difference must be made with the roles of engineers and of the closely related field of scientists.

Image courtesy of Tim Hunkin http://www.timhunkin.com, tim@timhunkin.com

The main difference between the engineer and scientist is in the object of each one's work. The scientist searches for answers to technological questions to obtain knowledge of why something occurs. The engineer

searches for answers to technological questions but always with an application in mind.
Theodore Von Karman, one of the pioneers of America's aerospace industry, said, "Scientists explore what is; engineers create what has not been." [Paul Wright, Introduction to Engineering].

In general, science is about discovering things or acquiring new knowledge. Scientists are always asking, "Why?" They are interested in advancing the knowledge base that we have in a specific area. The answers they seek may be abstract in nature, such as understanding the beginning of the universe, or may be more practical, such as the reaction of a virus to a new drug.

The engineer also asks why, but it is because a problem is preventing a product or service from being produced. The engineer is always thinking about the application when asking why. The engineer becomes concerned with issues such as the demand for a product, the cost of producing the product, the impact on society, and the environment of the product.

Scientists and engineers work in many of the same fields and industries but have different roles. Here are some examples:

- Scientists study the planets in our solar system to understand them; engineers study the planets so they can design a spacecraft to operate in the environment of that planet.

- Scientists study atomic structure to understand the nature of matter; engineers study the atomic structure to build smaller and faster microprocessors.

- Scientists study the human nervous system to better understand diseases; engineers study the human nervous system to design artificial limbs.

- Scientists create new chemical compounds in a laboratory; engineers create processes to mass-produce new chemical compounds for consumers.

- Scientists study the movement of the earth's tectonic plates to understand and predict earthquakes; engineers study the movement of tectonic plates to design safer buildings.

How Can We Save Water?

Content: Environment and conservation
Time: 1-2 class periods, 4 days at home collecting data.

Have you ever wondered how much water you use around the house? Many people never think about how much water they use in the performance of household chores. For example, some of the water in showering or washing dishes or brushing teeth is wasted. Water is a precious resource and its use should be carefully monitored. In this way, water will be conserved and there will be fresh water for everyone.

In this activity, you will examine and compare ways that your classmates use water as they brush their teeth. You will determine a more efficient way to brush your teeth and conserve water at the same time.

WALCH (1998) HANDS ON SCIENCE SERIES: Water. Pp. 47-50

THE ENGINEER AND THE ENGINEERING TECHNOLOGIST

Image courtesy of Teakoy http://www.teakoy.fi/engl/html/cnc.htm

Another profession closely related to engineering is engineering technology. Engineering technology and engineering have similarities, yet there are differences in that they have different career opportunities. ABET, which accredits engineering technology programs as well as engineering programs, defines engineering technology as follows:

Engineering technology is that part of the technological field, which requires the application of scientific and engineering knowledge and methods combined with technical skills in support of engineering activities. Engineering technology lies in the occupational range between the craftsman and engineering at the end of the range closest to the engineer.

Technologists work with existing technology to produce goods or "stuff" for society. Engineering technology students spend time in their studies working with actual machines and equipment that are used in the jobs they will accept after graduation. By doing this, technologists are equipped to be productive in their occupation from the first day of work.

Engineers and technologists apply technology for the betterment of society. The main difference between the two fields is that the engineer is able to create new technology through research, design, and development rather than being trained to use specific machines or processes.

There are areas where engineers and engineering technologists perform very similar jobs. For example, in manufacturing settings, engineers and technologists are supervisors of assembly line workers. Also, in technical service fields, both are hired to work as technical support personnel supporting equipment purchased by customers.

- The technologist identifies the computer networking equipment necessary for a business to meet its needs and oversees the installation of that equipment; the engineer designs new computer boards to transmit data faster.

- The technologist develops a procedure to manufacture a shaft for an aircraft engine using a newly developed welding technique; the engineer develops the new welding machine.

- The technologist analyzes a production line and identifies new robotic equipment to improve production; the engineer develops a computer simulation of the process to analyze the impact of the proposed equipment.

- The technologist identifies the equipment necessary to assemble a new CD player; the engineer designs the new CD player.

- The technologist identifies the proper building materials and oversees the construction of a new building; the engineer determines the proper support structures, taking into account the local soil, proposed usage, earthquake risks, and other design requirements.

WHAT DO ENGINEERS DO?

Within engineering are basic classifications of jobs common across the engineering disciplines. What follows are brief descriptions of these different engineering job functions. A few examples are provided for each function. All the fields of engineering have roles in each of the main functions described here.

Research

The role of the engineering researcher is the closest to that of a scientist of all the engineering functions. Research engineers explore fundamental principles of chemistry, physics, biology, and mathematics to overcome barriers preventing advancement in their field. Engineering researchers differ from scientists in that they are interested in the application of a breakthrough, whereas scientists are concerned with the knowledge that accompanies a breakthrough.

Research engineers conduct investigations to extend knowledge using various means. One of the means is conducting experiments. Research engineers may be involved in the design of experiments and the interpreting of the results. Typically, the research engineer does not perform the experiment. Instead, technicians usually do the testing. Large-scale experiments may involve the coordination of additional supporting personnel including other engineers, scientists, technologists, technicians, and craftspeople.

Dr. Steven Garrett, the United Technologies Corporation professor of acoustics at Penn State, with the new compact chiller prototype hooked up to a standard ice cream sales freezer. Dr. Garrett is also senior scientist at Penn State's Applied Research Laboratory Photo Credit: Penn State, Greg Grieco

Research is conducted using the computer. Computational techniques are developed to calculate solutions to complex problems without having to conduct costly and time-consuming experiments. Computational research requires the creation of mathematical models to simulate the natural occurring events under study. Research engineers might develop the computational techniques to perform the complex calculations in a timely and cost-effective fashion.

NASA's Glenn Research Center in Cleveland Ohio Image Courtesy of NASA

Most research engineers work for some type of research center. A research center might be a university, a government laboratory such as NASA, or an industrial research center. In most research positions, an advanced degree is required and, often, a Ph.D. is needed.

Development

Development engineers take the knowledge acquired by the researchers and apply it to a specific product or application. Development engineers bridge the gap between laboratory research and full-scale production. The development function is often coupled with research in research and development (R&D) divisions. The researcher may prove something is possible in a laboratory setting; the development engineer shows that it will work on a large, production-size scale and under field conditions. This is done in pilot manufacturing plants or by using prototypes.

Development engineers are always looking for ways to include researchers' findings into test products to test their likelihood for use in new products. Often, an idea proven in a laboratory needs to be significantly changed before it can be introduced on a large scale. The role of development engineers is to identify these areas and work with the design engineers to correct them before full-scale production begins. An example of a development process is the building of concept cars within the automotive industry. These unique cars include advanced design concepts and technology. The cars are then used as a test case to see if the design ideas and technology perform as predicted. The concept cars are put through exhaustive tests to determine how well the new ideas enhance a vehicle's performance. Each year, new technology is introduced into production automobiles that was first proven in development groups using concept vehicles.

Testing

Test engineers are responsible for designing and implementing tests to verify the truthfulness, reliability, and quality of products before they are

Race engineers look at data on overhead displays to set up the racecar and give information to the driver. Photo Credit: Al Gomez

introduced to the public. The test engineer devises ways to simulate the conditions a product will be subjected to during its life. Test engineers work closely with development engineers in evaluating prototypes and pilot facilities. Data from these initial development tests are used to decide if full production versions will be made or if significant changes are needed before a full-scale release. Test engineers work with design engineers to identify the changes in the product to ensure its truthfulness.

A challenge that engineers face is simulating the conditions a product will face during its life span and doing so in a timely, cost-effective manner. Often the conditions the product will face are difficult to simulate in a laboratory. A constant problem for the test engineer is simulating the product's aging. An example of such a testing challenge is the testing of a pacemaker, which is designed to last for several decades, for regulating a patient's heart. An effective test of this type cannot take 20 years or the product will be outdated before it is introduced. The test engineer must simulate conditions within the human body without exposing people to unnecessary risks.

Data acquisition engineers plug in their laptops to an open wheel racecar to pull data out of the car and provide information to the car's computer. Photo Credit: Al Gomez

Other challenges facing test engineers involve acquiring accurate and reliable data. The test engineer must produce data that show the product is functioning properly or identify areas of concern. Test engineers develop data acquisition and instrumentation methods to achieve this. Techniques such as radio telemetry may be used to transmit data from the inside of a device being tested. The measurement techniques must not interfere with the device's operation, presenting a tremendous challenge for small, compact products. Test engineers must cooperate with design engineers to determine how the tested device can be fitted with instrumentation yet still meet its design intent.

Test engineers must have a wide range of technical and problem-solving skills. They must be able to work in teams involving a wide range of people. They work with design and development engineers, technicians, and craftspeople, as well as management.

EXAMPLE 1

Test engineers must understand the important parameters of their tests. The development of a certain European high-speed train provides an example of the potential consequences, which may result when test engineers fail to understand these parameters. A test was needed to show that the windshield on the locomotive could withstand the high-velocity impacts of birds or other objects it might encounter. This common design constraint is encountered in airplane design. The train test engineers borrowed a "chicken gun" from an aerospace firm for this test. A chicken gun is a mechanism used to propel birds at a target, simulating in-flight impact. With modern laws governing cruelty to animals, the birds are humanely killed and frozen until the test.

An airline engineer fixes a planes window. Photo Credit: Al Gomez

On the day of the test, the test engineers aimed the gun at the locomotive windshield, inserted the bird, and fired. The bird not only shattered the windshield but put a hole through the engineer's seat.

The design engineers could not understand what went wrong. They checked their calculations and determined that the windshield should have held. The problem became clear after the test engineers reviewed their procedure with the engineers from the aerospace firm from whom they had borrowed the equipment. The aerospace test engineers asked how long they had let the bird thaw.

The response was, “Thaw the bird?”

There is a significant difference in the impact force between a frozen eight-pound bird and a thawed bird. The test was successfully completed later with a properly thawed bird.

DESIGN

The design function is what many people think of when they think of engineering, and this is where the largest number of engineers are employed. The design engineer is responsible for providing the detailed specifications of the products society uses.

Rather than being responsible for an entire product, most design engineers are responsible for a component or part of the product. The individual parts are then assembled into a product such as a computer, automobile, or airplane. Design engineers produce detailed dimensions and specifications of the part to ensure that it fits properly with adjoining pieces. They use modern computer design tools and are often supported by technicians trained in computer drafting software.

A design engineer would be responsible for an individual part in this telescope. Many engineers would put the telescope together.
Photo Credit: Richard Crisp

The form of the part is also a consideration for the design engineer. Design engineers use their knowledge of science and mathematical laws along with their experience to generate a shape to meet the part's specifications. Often, there is a wide range of possibilities and considerations. In some fields, these considerations are ones that can be calculated. In others, such as in consumer products, the reaction of a potential customer to a shape may be as important as how well the product works.

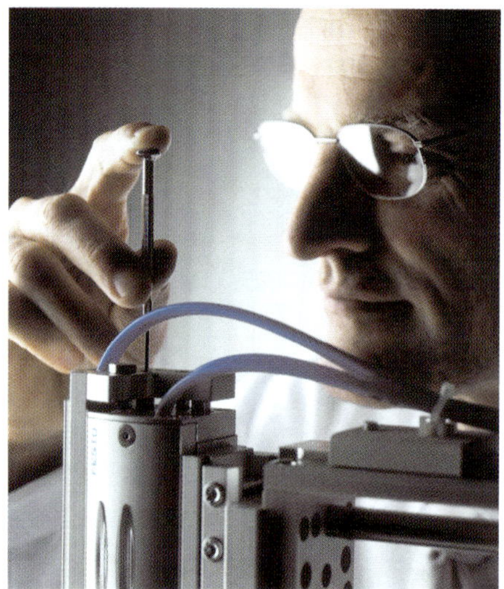

Image courtesy of Wordsun.com

The design engineer must verify that the part meets the reliability and safety standards established for the product. This often requires coordination with other engineers to simulate complex products and field conditions. The design engineer is responsible for making corrections to the design based on the results of the tests performed.

In today's world of ever-increasing competition, the design engineer must involve manufacturing engineers in the design process. Cost is a critical factor in the design process and may be the difference between a successful product and one that fails. Communication with manufacturing engineers is therefore, critical. Often, simple design changes can radically change a part's cost and affect the ease with which the part is made.

Design engineers work with existing products. Their role includes redesigning parts to reduce manufacturing costs and time. They work on redesigning products that have not lived up to expected lives or have suffered failure while being used. They modify products for new uses. This usually requires additional study, minor redesigns, and significant communication between departments.

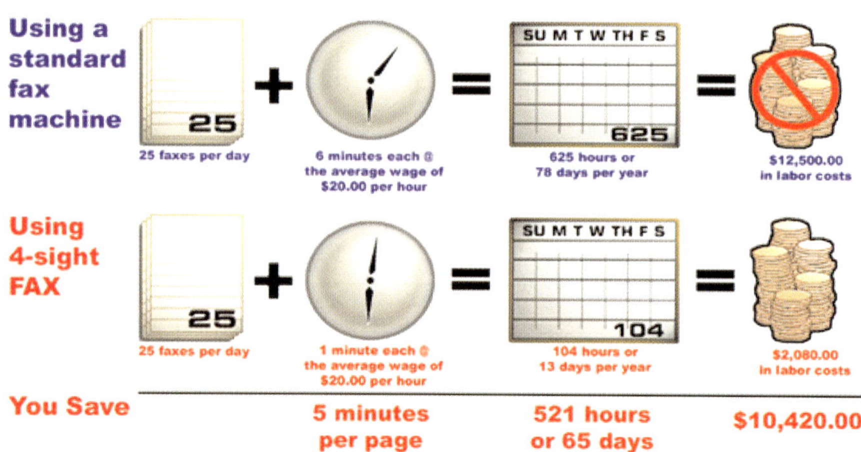

A sample cost analysis diagram. Image courtesy of 4sightfax.com

IDEAS Alternative Energy-Windmills

The machine may be designed to transport a given load (weight in marbles) for the furthest distance within a given amount of time. The success of the invention may be measured by calculating the efficiency of the machine by using the following formula(s):

Power in = # watts = power rating of fan
Power out = (mgL)/time

where:
m = mass of marbles (kg)
g = gravity coefficient 9.81 m/sec2
L = the distance (length) between tape marks on string
Time = time from mark 1 to mark 2

Students may only use the materials provided to design and make the wind machine. The suggested completion time is five class periods of 40 minutes each. During that time, the students may trial test the machine at any time, but should only be given three attempts to achieve the greatest efficiency; this will permit them to make revisions to their original designs.

PROBLEM-SOLVING PROCESS:

These steps may be helpful to students in approaching their activity:

- Form cooperative groups (2 to 3 people)
- Brainstorm for ideas
- Sketch possible solutions
- Decide how to construct, maneuver, operate, etc. the project
- Select and gather materials
- Construct your design
- Test your design
- Present your design

CASE STUDY

The Earth has provided people with many resources for energy throughout history. Though many of these resources have been used extensively, other excellent sources of power have been ignored. The wind is one of these overlooked energy sources.

Everyone has witnessed the positive and negative effects of moving air or wind on our structures and the environment. Human beings have attempted to harness this invisible power to make their lives easier for nearly 5,000 years. In Egypt, wind was first used to propel watercraft. In ancient Babylon, stationary wind devices were used to convert wind into useful forms of energy and motion. Wind turbines and windmills later emerged throughout Asia and Europe. Since the 1300's, the Dutch have used windmills to drain land that lies below sea level. Farmers throughout the world have used windmills to pump ground water to feed livestock and to sustain crops. In the early nineteenth century, fossil fuels such as coal and oil became the standard source of energy. With the invention of the steam engine, the use of wind power and other alternate sources declined.

Today, there is renewed interest in wind power because of the undesirable environmental effects of fossil fuel power. Wind has been an ineffective economic source of energy, but new methods are making wind power a more efficient, competitive, and promising source of clean alternative energy. It is becoming more apparent that alternative energies may be essential and not just desirable. Many improvements have already been made, and new ways of packaging, shipping, and recycling can make even more of a difference. Our next task is to control our energy consumption. Can you think of other sources of clean and natural energy? How are these other energies used?

CHALLENGE

EXPLORATORY LEVEL

Research and model a historical wind machine. Display how a person's life was made easier because the machine was available.

INTERMEDIATE LEVEL

Design and model a wind-powered sculpture that is appealing to the eye, is environmentally friendly, and incorporates some form of sound.

Design and make a safe toy for a young child that is powered by the wind. Your design should be some type of mechanism that incorporates movement and color.

ADVANCED LEVEL

Design and make a wind machine that utilizes the wind as an energy source (e.g., an electric fan) and is capable of producing work (moving a given load). The machine should be designed to transport a given load (using marbles) for the furthest distance within a given amount of time. The success of the invention may be measured by calculating the efficiency of the machine.

MATERIALS:

EXPLORATORY	INTERMEDIATE	ADVANCED
Cardboard	Cardboard	Cardboard
Paper, various weights/colors	Paper, various weights/colors	Paper 8.5" x 11"
Masking tape	Masking tape	Masking tape
Drinking straws	Drinking straws	Drinking straws
Craft sticks/Balsa wood	Paint, crayons, markers	Drinking cups
Kite string	Rubber bands	Straight pins
Other found materials	Craft sticks/Balsa wood	Pencil with eraser
Scissors, glue, fasteners, etc.	Kite string	Kite string
	Found materials (cans, bottles, recycled items)	Stopwatch
	Bells, noise makers, etc.	Scissors or X-acto knife
	Scissors, glue, fasteners, etc.	Glue
		Marbles

ASSESSMENT

Your success on this Challenge will be based on your completion of the activities below. Three general criteria for your performance will be your participation in the activity, the accuracy of your measurement and model construction, and the performance of your design. Your teacher will help you understand how your performance will be graded.

EXPLORATORY

1. Locate and document at least three examples of wind machines from the past.
2. Write a brief paragraph that summarizes the benefits that each wind machine provided to human beings during the specific time period.
3. Choose one of the historical wind machines and provide an explanation of its operation.
4. Construct a model of the chosen wind machine demonstrating how it made life easier.
5. Test and present your design.

INTERMEDIATE

ONE:

1. Research windmills and other wind-powered machines.
2. Research sculptures, mobiles, and other three-dimensional art.
3. Gather found materials, recycled items, and sound devices (bells, noise makers).
4. Draw several sketches of a wind-powered sculpture. Choose the best design and illustrate what it will look like in a garden or park environment.
5. Construct your design using only the materials provided.
6. Test and present your design.

TWO:

1. Research wind powered machines and vehicles.
2. Brainstorm ideas for a wind-powered toy, safe for a child.
3. Sketch several plans that will use color and movement effectively.
4. Design your project; gather necessary materials, construct the project.
5. Test and present your design.

ADVANCED

1. Research wind-powered machines, structures and other devices affected by wind.
2. Research and review simple machines and how they help to make life easier.
3. Brainstorm ideas to use wind power to transfer energy so a given load may be lifted, pushed, or pulled from one point to another.
4. Use those ideas to design and construct a wind-driven vehicle. Decide upon a set of materials that all participants must use. A fan can be the wind source.
5. Test and redesign, if necessary, so the greatest load will be transported the greatest distance in any given amount of time.

Resources

- http://www.surfnetkids.com/wind.htm
- http://www.igc.apc.org/awea/aweanews.html (provided by "WIND NEWS")
- http://solstice.crest.org/renewables/re-kiosk/wind/index.shtml
- Amazon Books: ("Power from Wind: A History of Windmill Technology" by Richard L. Hills ISBN: 052156686X; "Wind Energy in America" by Robert W. Righter ISBN: 0806128127; and "Wind Power - Energy Forever" by Ian Graham to be published 1999 ISBN: 0817253645)

IDEAS is the result of a project funded by The Engineering Foundation and organized by two of the worlds major engineering societies, the American Society of Mechanical Engineers (ASME) and the American Society of Civil Engineers (ASCE). Great lakes press has been grated permission by ASME and ASCE for the use of IDEAS projects within this textbook

ANALYSIS

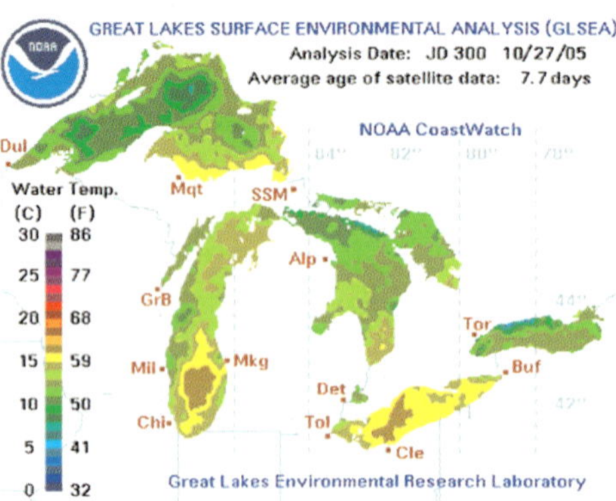

Surface temperature for the great lakes of the United States. This image is courtesy of NOAA.org

Analysis is an engineering function performed in conjunction with design, development, and research. Analysis engineers use mathematical models and computational tools to provide the necessary information to design, development, or research engineers to help them perform their function.

Analysis engineers typically are specialists in a technology area important to the products or services being produced. Technical areas might include heat transfer, fluid flow, vibrations, dynamics, system modeling, and acoustics. They work with computer models of products to make these assessments. Analysis engineers often possess an advanced degree and are experienced in their area of expertise.

To produce the information required of them, they must confirm their computer programs or mathematical models. To do so may require comparing test data to their predictions. This requires coordination with test engineers to design an appropriate test and to record the relevant data.

An example of the role of analysis is the prediction of temperatures in an aircraft engine. Material selection, component life estimates, and design decisions are based in large part on the temperature the parts reach and the duration of those temperatures. Heat transfer analyses are used to determine these temperatures. Engine test results are used to confirm the temperature predictions. The permissible time between engine overhauls can depend on these temperatures. The design engineers then use these results to ensure reliable aircraft propulsion systems.

A race team owner and general manager go over racecar data on the monitors. Photo Credit: Al Gomez

SYSTEMS

Systems engineers work with the overall design, development, manufacture and operation of a complete system or product. Design engineers are involved in the design of individual components, but systems engineers are responsible for the integration of the components and systems into a functioning product.

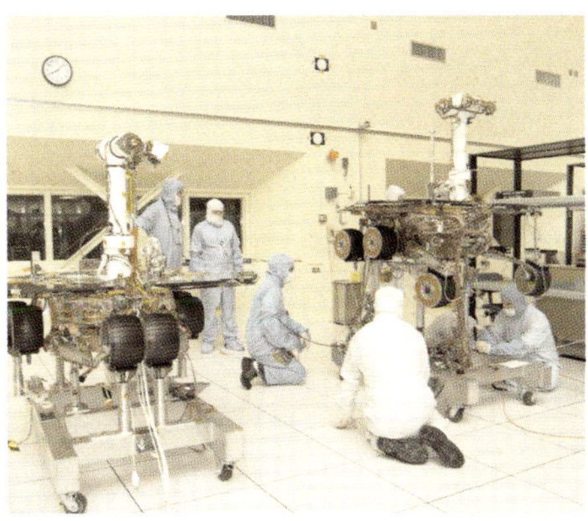

Mars rovers' engineers put the final touches on the robots that are now on Mars surveying the planet. Image courtesy of NASA

Systems engineers are responsible for ensuring the components work together properly and work as a complete unit. Systems engineers are responsible for identifying the overall design requirements. This may involve working with customers or marketing personnel to determine market needs. From a technical standpoint, systems engineers are responsible for meeting the overall design requirements.

Systems engineering is a field that most engineers enter only after becoming proficient in an area important to the systems, such as component design or development.

MANUFACTURING AND CONSTRUCTION

Manufacturing engineers turn the specifications of the design engineer into a reality. They develop processes to make the products we use every day. They work with diverse teams of individuals, from technicians on the assembly lines to management, to maintain the truthfulness and efficiency of the manufacturing process.

It is the responsibility of manufacturing engineers to develop the processes for taking raw materials and changing them into the finished pieces that the design engineers detailed. They utilize the best machines and processes to accomplish this. As technology advances, new processes often must be developed for manufacturing the products.

The repeatability or quality of manufacturing processes is an area of increasing concern to modern manufacturing engineers. These engineers use statistical methods to determine the precision of a process. This is important since a lack of precision in the manufacturing process may result in low-grade parts that cannot be used or that may not meet the customer's needs. Manufacturing engineers are concerned about the quality of the products they produce. High-quality manufacturing means lower

Engineers work together on a robotic work cell. Image courtesy of Weber State Universities department of Manufacturing Engineering Technology department.

costs since more parts are usable and less is wasted. Ensuring quality means having the right processes in place, understanding the processes, and working with the people involved to make sure the processes are being maintained at peak efficiency.

Manufacturing engineers keep track of the equipment in a plant. They schedule and track required maintenance to keep the production line moving. They track the inventories of raw materials, partially finished parts, and completely finished parts. Excessive inventories tie up substantial amounts of cash that could be used in other parts of the company. The manufacturing engineer is responsible for maintaining a safe and reliable workplace, including the safety of the workers at the facility and the environmental impact of the processes.

Manufacturing engineers must be able to work with diverse teams of people including design engineers, tradesmen, and management. Current "just in time" manufacturing practices reduce needed inventories in

Engineers have to work together to solve problems and make modifications on projects. Image courtesy of Georgian College

factories but require manufacturing engineers at one facility to coordinate their operation with manufacturing engineers at other facilities. They must coordinate the line workers who operate the manufacturing equipment. Manufacturing engineers must maintain a constructive relationship with their company's trade workers.

Manufacturing engineers play a critical role in modern design practices. Since manufacturing costs are such an important component in the success of a product, the design process must take into account manufacturing concerns. Manufacturing engineers identify high-cost or high-risk operations in the product's design phase. When problems are identified, they work with the design engineers to generate alternatives.

In the production of large items such as buildings, dams, and roads, the production engineer is called a construction engineer rather than a manufacturing engineer. However, the construction engineer's role is similar to that of the manufacturing engineer. The main difference is that the construction engineer's production facility is typically outdoors, but the manufacturing engineer's is inside a factory. The functions of construction engineers are the same as mentioned above when "assembly line" and "factory" are replaced with terms like "job site," reflecting the construction of a building, dam, or other large-scale project.

CASE STUDY

An automated crucible pours a molten metal into a mold. Image courtesy of the European Industrial Minerals Association

Casting has been around for years and is used to make parts for cars, video games, sporting equipment, and yes, for making candy. Jewelers have used casting to make their designs, automobile manufacturing companies cast their parts for their cars, and sports companies use casting to make protective equipment for players.

You will be given time from your instructor to cast your chocolate mold and then weigh it, recording this information to compare to the class later. The objective is to create the most "perfect" casting by not using too much or little chocolate to fill the mold. Your weight will be compared to the weight of a perfect mold and to your classmates' molds. The data will be compiled by your instructor on the board so you can see how many different weights occurred.

In addition to the activity, answer these three questions:

1. If you were a candy maker, would you want to have more chocolate in the mold than it can hold?
2. How much variation was there between your mold and the rest of the class?
3. How could you control the amount of chocolate that goes into the mold?

OPERATIONS AND MAINTENANCE

After a new production facility is brought on-line, it must be maintained. The operations engineer oversees the ongoing performance of the facility. Operations engineers must have a wide range of expertise dealing with the mechanical and electrical issues involved with maintaining a production line. They must be able to interact with manufacturing engineers, line workers, and technicians who service the equipment. They must coordinate the service schedule of the technicians to ensure efficient service of the machinery, minimizing its downtime impact on production.

A large production facility focuses on mining gold. Image courtesy of Copeland Industries

Maintenance and operations engineers work in non-manufacturing roles. Airlines have staffs of maintenance engineers who schedule and oversee safety inspections and repairs. These engineers must have expertise in sophisticated inspection techniques to identify all possible problems.

Large medical facilities and other service sector businesses require operations or maintenance engineers to oversee their equipment's operation. Emergency medical equipment must be maintained in peak working order.

TECHNICAL SUPPORT

A technical support engineer serves as the link between customer and product and assists with installation and setup. For large industrial purchases, technical support may be included in the purchase price. The engineer may visit the installation site and oversee a successful startup. For example, a new power station for moving water into farmers' fields might require the technical support engineer to supervise the installation and help the customer solve problems to get the product operational. To be effective, the engineer must have good interpersonal and problem-solving skills as well as solid technical training.

A computer help desk is a great place to call to get technical support for your problems. Image courtesy of Virginia Military Institute.

The technical support engineer may troubleshoot a product's problems. Serving on a computer company's help line is one example. Diagnosing design flaws found in the field once the product is in use is another example.

Technical support engineers do not have to have in-depth knowledge of the product. However, they must know how to tap into such knowledge at their company.

Modern technical support is being used as an added service. Technical support engineers work with customers to operate and manage their own company's equipment as well as others. For example, a medical equipment manufacturer might sell its services to a hospital to manage and operate its highly sophisticated equipment. The manufacturer's engineers would maintain the equipment and help the hospital use its facilities in the most efficient way.

SALES

Engineers are valuable members of the sales force in numerous companies. These engineers must have interpersonal skills conducive to effective selling. Sales engineers bring many assets to their positions.

Engineers have the technical background to answer customer questions and concerns. They are trained to identify which products are right for the customer and how they can be applied. Sales engineers can identify other applications or other products that might benefit the customers once they become familiar with their customer's needs.

In some sales forces, engineers are used because the customers are engineers and have engineering-related questions. When airplane manufacturers market their aircraft to airlines, they send engineers. The airlines have engineers who have technical concerns overseeing the aircraft's maintenance and operation. Sales engineers have the technical background to answer these questions.

As technology advances, more products will become technically sophisticated. This produces an increasing demand for sales engineers.

CONSULTING

Consulting engineers are self-employed or they work for a firm that does not provide goods or services directly to consumers. Such firms provide technical expertise to organizations that do. Many large companies do not have technical experts on staff in all areas of operation. Instead, they use consultants to handle issues in those technical areas.

For example, a manufacturing facility in which a cooling tower is used in part of the operation might have engineers who are well versed in the manufacturing processes but not in cooling tower design. The manufacturer would hire a consulting firm to design the cooling tower and related systems. Such a consultant might oversee the installation of such a system or provide a report with recommendations. After the system is in place or the report is delivered, the consultant would move on to another project with another company.

Consulting engineers evaluate the effectiveness of an organization. In such a situation, a team of consultants might work with a customer and provide suggestions and guidelines for improving the company's processes. These might be design methods, manufacturing operations, or business practices. Though some consulting firms provide only engineering-related expertise, other firms provide engineering and business support and require the consulting engineers to work on business-related issues as well as technical issues.

Consulting engineers interact with a wide range of companies on a broad scope of projects and come from all engineering disciplines. Often, a consultant needs to be registered as a professional engineer in the state where he or she does business.

MANAGEMENT

In many instances, engineers move into project management positions and into full-time management. National surveys show that more than half of all engineers will be involved in some type of management responsibilities, supervisory or administrative, before their career is over. Engineers are chosen for their technical ability, problem-solving ability, and leadership skills.

Engineers may manage other engineers or support personnel or they may rise to oversee the business aspects of a corporation. Prior to being promoted to this level of management, engineers may acquire some business or management training. Some companies provide this training or offer incentives for employees to take management courses in the evening on their own time.

OTHER FIELDS

Some engineering graduates enter fields other than engineering, such as law, education, medicine and business. Patent law is one area in which an engineering or science degree is almost essential. In patent law, lawyers research, write, and file patent applications. Patents grant an inventor the right to exclude others from producing or using the inventor's discovery or invention for a limited period of time. Patent lawyers must have the technical background to understand what an invention does so they can describe the invention and legally protect it for the inventor.

Another area of law that has become popular for engineering graduates is corporate liability law. Corporations are faced with decisions every day over whether to introduce a new product, and they must weigh the potential risks. Lawyers with technical backgrounds have the ability to weigh technical as well as legal risks. Such lawyers are used in litigation for liability issues. Understanding what an expert witness is saying in a suit can be a tremendous advantage in a courtroom and can enable a lawyer to cross-examine the expert witness effectively. Often, when a corporation is sued over product liability, the lawyers defending the corporation have technical backgrounds.

Engineers are involved in several aspects of education. The one that students are most familiar with is an engineering professor. College professors usually have PhDs. Engineers with master's degrees can teach at community colleges and in some engineering technology programs. Engineering graduates can teach in high schools, middle schools, and elementary schools. To do this full-time usually requires additional training in educational methods, but the engineering background is a great start. Thousands of engineers are involved in part-time educational projects where they go to classes as guest speakers to show students what "real engineers" do (See example 2 below). The American Society for Engineering Education (ASEE) is a great resource if you are interested in engineering education.

Engineers find careers in medicine, business, on Wall Street, and in many other professions. In modern society, with technology's rapid expansion, the combination of problem-solving ability and technical knowledge makes engineers valuable and versatile.

EXAMPLE 2

Listed below are things that a mechanical engineer (ME) might perform in a design function. This illustrates how a specific job in one industrial sector can vary from that in another sector. Everything listed below is what a mechanical engineer might do in the specified field.

- **Aerospace**. Design of an aircraft engine fan blade: Detailed computer analyses and on-engine testing are required for certification by the Federal Aviation Administration. Reliability, efficiency, cost, and weight are all design constraints. The design engineer must push current design barriers to optimize the design constraints, potentially making tradeoffs between efficiency, cost, and weight.
- **Biomedical**. Design of an artificial leg and foot prosthesis giving additional mobility and control to the patient: Computer modeling is used to model the structure of the prosthesis and the natural movement of a leg and foot. Collaboration with medical personnel to understand the needs of the patient is a critical part of the design process. Reliability, durability, and functionality are the key design constraints.

- **Power**. Design of a heat recovery system in a power plant, increasing the plant productivity: Computer analyses are performed as part of the design process. Cost, efficiency, and reliability are the main design constraints. Key mechanical components can be designed with large factors of safety since weight is not a design concern.
- **Consumer products**. Design of a pump for toothpaste for easier dispensing: Much of the development work might be done with prototypes. Cost is a main design consideration. Consumer appeal is another consideration and necessitates extensive consumer testing as part of the development process.
- **Computer**. Design of a new ink-jet nozzle with resolution approaching laser printer quality: Computer analyses are performed to ensure that the ink application is properly modeled. Functionality, reliability, and cost are key design concerns.

Chapter 2

History of Engineering

Introduction

Studying history provides the best way to know where we have been, where we are, and where we are going.

There were few engineering innovations in the early years of man. However, as time passed, innovations occurred more quickly. Today, engineering discoveries are made almost daily. The speed with which things change indicates the urgency of understanding the process of innovation.

History provides a great opportunity to observe the context of before, during, and after of some of the greatest engineering problems ever faced, and of those of today.

History is not about memorizing names and dates. History is about people. The information that can be learned from stories about how engineers developed products can be motivating, informative, and helpful. These stories provide the foundation needed to learn and to design in the future.

Many types of professionals are required to study the history of their trade as part of their education. Professional development can be a significant factor in determining who moves forward in their careers and who doesn't. The best professionals keep studying throughout their ca-

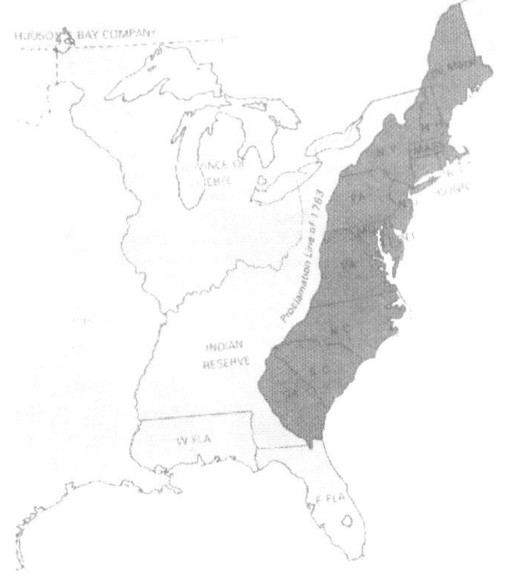

A map of the USA in 1775 From the essay "The Battle of Lexington" By Brenda Kamphuis

reers. Reading published papers, professional magazines, and journals builds a knowledge of history.

The study of history helps us create new futures and helps us understand what good qualities from the past are worth maintaining. Craftsmanship, integrity, and dedication are evident in engineering artifacts. History is also full of interesting, educational stories, characters, and clever developments.

The engineer has always affected the human race at every stage of its development. The few items mentioned in this chapter are only highlights of contributions that engineers have made to the progress of humanity.

Prehistoric Culture

Prehistoric people practiced engineering that we might label as engineering problem solving, tool-making, etc., but they did not have a grasp of mathematical principles or the knowledge of natural science as we know it today. These cultures and civilizations designed and built needed items through trial and error and as ideas simply that occured to them. They built some spears that worked and some that failed, but in the end, they perfected weapons that allowed them to kill game animals and to feed their families. Because there was little written communication and only primitive transportation, little information and few ideas were exchanged with people from other places. Each civilization inched ahead on its own.

However, some of these prehistoric innovators would have made fine engineers today. Even with their limited skills, their knowledge of their surroundings was more extensive than modern civilization typically can comprehend. Their craftsmanship was often marvelous in its effectiveness, reliability, and sophistication. They passed on knowledge of all aspects of life to the next generation. This information was carefully memorized and kept accurate, growing, and alive. Early man even tried to pass on vital information by way of coded cave paintings and etchings.

The physical limitations of prehistoric man can perhaps be highlighted as follows:

- They had no known written language.
- Their verbal language was limited.
- They had no means of transportation.
- They had no separate concept of education or specialized methodology to discover new things.
- They lived by gathering food and killing game with primitive weapons.
- Improvement of the material aspects of life came about slowly, with early, primitive engineering.

We will present a panoramic view of engineering by briefly stating some of the more interesting happenings during specific time periods. Notice the kinds of innovations that were introduced. Examine the relationships many inventions had with each other. Think about the present and the connectivity between all areas of engineering and the critical importance of the computer. Innovations do not happen in a vacuum; they are interrelated with the needs and circumstances of the surrounding world.

The Rate of Advancement in History

The rate of innovation brings up some interesting points. As we move through the past 6,000 years, you should realize that the rate at which we currently introduce innovations is far more rapid than in the past. It used to take years to accomplish tasks that today we perform in a short time. In the past, there were often decades without noticeable technological progress. Think of the amount of time that it takes to construct a building today with the equipment that has been developed by engineers. A complete house-frame can be constructed in a single day now, yet Churches still stand today that took as long as 200 years to construct when they were built.

The ancient people accomplished fantastic feats with only a basic knowledge of the principles that we learned as children. Some of this knowledge and the speed with which it was introduced or practiced may have been purposely held back. Archimedes refused to release information that could have been used to make more effective weapons; he knew it would be used for harm and not for the pursuit of wisdom. Only when his home city of Syracuse was no longer able to hold off the Roman attackers did he release his inventions to the military. The study of history confronts us with dilemmas. Neither side of a true dilemma ever goes away. The story of history never ends.

Our technological roots can be traced back to the gatherer/hunter. Prehistoric people survived by collecting food and killing what animals they could chase down. They endured a lean physical existence. As they improved their security through innovation, humanity increased in numbers, and they had to find ways to feed and to control their growing population. To support larger populations reliably, the methods and implements of farming and security were improved. Much later, after many smaller innovations, specialized industrial cultures stepped onto the scene, ready to bring the world productivity and material wealth.

THE BEGINNINGS OF ENGINEERING

The Earliest Days

The foundations of engineering were laid with our ancestor's efforts to survive and to improve their quality of life. From the beginning they looked around their environment and saw areas where life could be made easier and more stable. They developed improved ways to hunt and to fish. They discovered better methods for providing shelter for their families. Their main physical concern was daily survival. As life became more complicated and small collections of families became larger communities, the need grew to look into new areas of concern, for example, power struggles, acquisition of neighboring tribes' lands, and religious observances. All of these involved work with tools. Engineering innovations were needed to further these interests. Of course, in those days projects were not thought of as separate from the rest of life. In fact, individuals were not generally thought of as being separate from their community. They did not look at life from the point of view of specialties and individual interests. Every person was an engineer, to an extent.

There are pockets of people that still live today much as their ancestors did in prehistoric times. However, frequently, even they take advantage of modern engineering in the form of tools, motors, and medicine. The Amish fall into this category of being a roots-oriented culture. Such cultures tend to use tools only for physical necessities so the significance of objects does not pollute their way of life.

Photo Credit: Robin Kravets, Department of Computer Science, University of Illinois, Urbana-Champaign

Egypt and Mesopotamia

As cities grew and the need for addressing the demands of fledgling societies increased, a significant change took place. People who showed special aptitude in certain areas were identified and assigned to more specialized tasks. This labeling and grouping was a scientific breakthrough. It gave toolmakers the time and resources to dedicate themselves to building and innovating. This new social function created the first real engineers, and for the first time, innovation flourished rapidly.

Between 4000 and 2000 B.C., Egypt and Mesopotamia were the focal points for engineering activity. Stone tools were developed to help in the quest for food. Copper and bronze axes were perfected through smelting. These developments were aimed not only at hunting. The development of the plow allowed man to become a farmer so he could reside in one place, and leave the nomadic way of life. Mesopotamia made its mark in the field of engineering by developing the wheel, the sailing boat, and writing methods. Engineering skills that were applied to the development of everyday items immediately improved life as they knew it. We will never completely understand the importance of the Greeks, Romans, and Egyptians in the life of the engineer.

During the construction of the Pyramids (c.2700-500 B.C.), the number of engineers required was immense. They had to make sure that everything fit correctly, that stones were properly transported long distances, and that the tombs would be secure against robbery. Imhotep (chief engineer to King Zoser) was building the stepped pyramid at Sakkara, in Egypt, about 2700 B.C. The more elaborate Great Pyramid of Khufu (all pictures) would come about 200 years later. By investigating the construction of the Pyramids, we learn about the need for designing, building, and testing with any engineering project. These early engineers, using only simple tools, performed with great capability, insight, and technical sophistication, tasks that even today give us a sense of pride in their achievements.

The Great Pyramid of Khufu is the largest masonry structure ever built. Its base measures 756 feet on each side. The 480-foot tall structure was constructed of over 2.3 million limestone blocks with a total weight of over 58,000,000 tons. Casing blocks of fine limestone were attached to all four sides. These casing stones, some weighing as much as 15 tons, have been removed over the centuries for a wide variety of other uses. It is hard for us to imagine the engineering expertise needed to quarry and move these base and casing stones, and then piece them together so that they would form the pyramid and its covering.

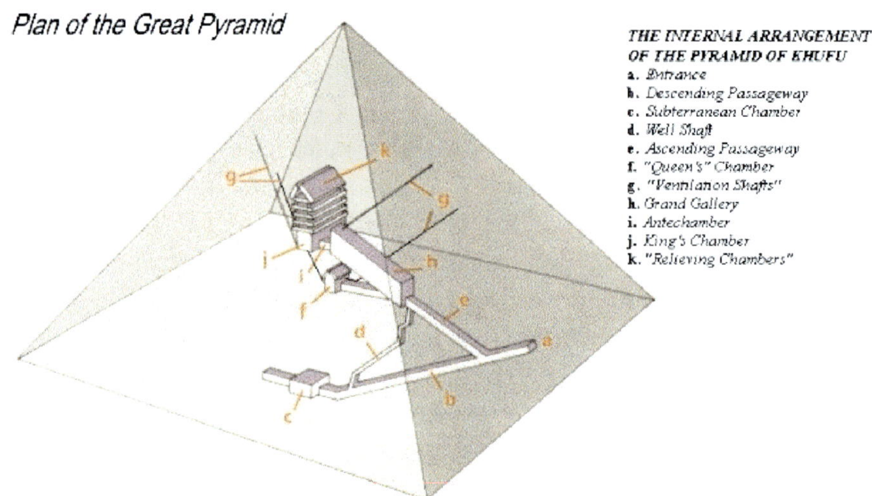

The Secret Doors Inside the Great Pyramid (copyright Dr. Zahi Hawass)

Here are additional details about this pyramid given by Roland Turner and Steven Goulden in Great Engineers and Pioneers in Technology, Volume 1: From Antiquity through the Industrial Revolution:

> "Buried within the pyramid are passageways leading to a number of funeral chambers, only one of which was actually used to house Khufu's remains. The granite-lined King's Chamber, measuring 17 by 34 feet, is roofed with nine slabs of granite which weigh 50 tons each. To relieve the weight on this roof, located 300 feet below the apex of the pyramid, the builder stacked five hollow chambers at short intervals above it. Four of the "relieving chambers" are roofed with granite lintels, and the topmost has a corbelled roof. Although somewhat rough and ready in design and execution, the system effectively distributes the massive overlying weight to the sturdy walls of the King's Chamber.
>
> Precision marks every other aspect of the pyramid's construction. The four sides of the base are practically identical in length, the error is a matter of inches and the angles are equally accurate. Direct measurement from corner to corner must have been difficult since the pyramid was built on the site of a rocky knoll (now completely enclosed in the structure). Moreover, it is an open question how the builder managed to align the pyramid almost exactly north-south. Still, many of the techniques used for raising the pyramid can be deduced.

> After the base and every successive course was in place, it was leveled by flooding the surface with Nile water, no doubt retained by mud banks, and then marking reference points of equal depth to guide the final dressing. Complications were caused by the use of blocks of different heights in the same course."

The above excerpt mentions a few of the fascinating details of the monumental job undertaken to construct a pyramid with only primitive tools and human labor.

This entrance to the Great Pyramid of Khufu (Cheops) leads to the pharaoh's simple funeral chamber. Photo Credit: Michael A. Stecker

IDEAS: CATAPULT

PROBLEM SOLVING PROCESS:

These steps may be helpful to students in approaching their activity:

- Form cooperative groups (2 to 3 people)
- Brainstorm for ideas
- Sketch possible solutions
- Decide how to construct, maneuver, operate, etc., the project
- Decide on and gather materials
- Construct your design
- Test your design
- Present your design

CASE STUDY

Throughout the history of the world, humankind has endured numerous destructive wars. While war was never a desirable choice for civilizations, societies have always felt compelled to be prepared for an attack. This constant pursuit of preparedness has produced numerous technological advances. One such innovation was the catapult.

The ancient Greeks and Romans were the first to perfect this device. Their catapults were capable of throwing a 60-pound rock farther than the distance of five football fields. It took great ingenuity and skill to design and build a device that harnessed such a tremendous amount of energy.

Modern catapults are based upon design principles derived from the early Greek and Roman catapults. For example, the massive steam catapults used to launch aircraft from carriers at sea are adaptations of these principles. By providing a rapid controlled energy source and the right angle of launch, objects can be thrown long distances with amazing accuracy.

CHALLENGE

EXPLORATORY

Research and model a historical catapult that is capable of launching a penny at least 10 meters.

INTERMEDIATE

Research and construct a model of a working catapult. Your design must be able to launch a load (Ping-Pong ball) from a tabletop onto a 600mm diameter target (box) placed at least three (3) meters away on the floor.

ADVANCED

Research, design and construct a catapult that will launch various loads (pennies, etc.) from a table top into a 15cm diameter container located five (5) meters away on the floor.

MATERIALS:

EXPLORATORY, INTERMEDIATE, ADVANCED

- Cardboard
- Paper of various weights/colors
- Plastic spoons
- Rulers/Tape measures
- Masking tape
- Rubber bands
- Craft sticks/Balsa wood
- Kite string
- Scissors, glue, fasteners
- Box for target
- 1-land saws/Drills/Wood files
- Finishing nails
- Straight pins
- X-acto knife
- Protractors
- Compass
- Pennies

- Ping-Pong ball
- 15cm diameter container
- Additional found objects and/or materials from home
- Safety Glasses

CRITERIA:

Your success on this Challenge will be based on your completion of the activities below. Three general criteria for your performance will be your participation in the activity, the accuracy of your measurement and model construction, and the performance of your design. Your teacher will help you understand how your performance will be graded.

EXPLORATORY

- Locate and document several examples of catapults from the past.
- Write a brief paragraph that summarizes the historical contribution(s) to civilizations that catapults provided.
- Choose a historical catapult model and develop an explanation of the mechanics of its operation.
- Construct a working model of the selected catapult and include sketches and drawings.
- Test and present your design.

INTERMEDIATE

- Research catapults and other war machines from the past.
- Summarize the history and mechanics of a specific catapult.
- Construct a working model of the chosen catapult using only the materials provided. Be sure to include sketches and drawings.
- Conduct tests to acquire Performance Data (range and accuracy) and chart the results.
- Test the accuracy of the design and predict the range of the evaluation toss.
- Present your design.

ADVANCED

- Research catapults and other war machines from the past.
- Summarize the history and mechanics of different catapults.
- Develop a design that provides an adjustable angle of launch and a triggering mechanism.

- Construct a working model of the chosen catapult using only the materials and specifications provided. Be sure to include sketches and drawings.
- Conduct tests to acquire Performance Data (maximum height of launch, maximum range and accuracy) and chart the results.
- Use the Performance Data to predict the distance of each launch. Compare the prediction to results.
- Present your design.

Resources

"History of Technology," London: Mansell Publishing
TIES Magazine, http://www.tiesmagazine.org
Various Ancient and World History Texts
National Catapult Museum, www.nzp.com

IDEAS is the result of a project funded by The Engineering Foundation and organized by two of the worlds major engineering societies, the American Society of Mechanical Engineers (ASME) and the American Society of Civil Engineers (ASCE). Great lakes press has been granted permission by ASME and ASCE for the use of IDEAS projects within this textbook.

Outline of engineering history

As we enter the modern age, invention appears to narrow, with much of the activity relating to computers. What is the role of the inventor in history? Sometimes the name is important and other times not. During certain eras, innovations were being made at the same time by a number of people, so no individual stands out. It might have been that the developments were more collective during such times. The people involved were often racing each other to the patent office. At other times, with certain inventions, a single person made a significant breakthrough on his or her own.

When is it the person and when is it the times? Here are a few clues. When it's the person, peers might think of him or her as crazy and might outlaw or ignore the work. When it's the times, many innovators are frequently doing similar work in close proximity or in cooperation. Sometimes when a name stands out, the impression remains that his or her effort was more communal than singular. These are concepts to consider as time and innovations flow past in these lists and in future studies.

1200 B.C. - A.D. 1

- Swords are mass produced.
- Siege towers are perfected.
- The Greeks develop manufacturing.
- Archimedes introduces mathematics in Greece.
- Concrete is used for the arched bridges, roads, and aqueducts in Rome.

A.D. 1 - 1000

Photo Credit: Jean Luc Eliss

- The Chinese further develop the study of mathematics.
- Gun powder is perfected.
- Cotton and silk are manufactured

1000 - 1400

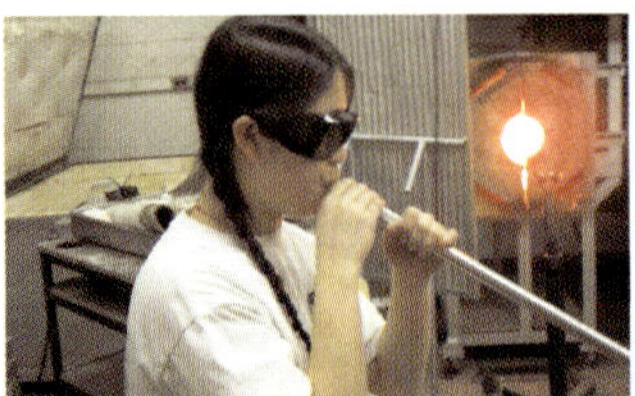

Photo Credit: Jason Gallicchio

- The silk and glass industries expand.
- Leonardo Fibonacci (1170-1240), medieval mathematician, writes the first Western text on algebra.

1400 - 1700

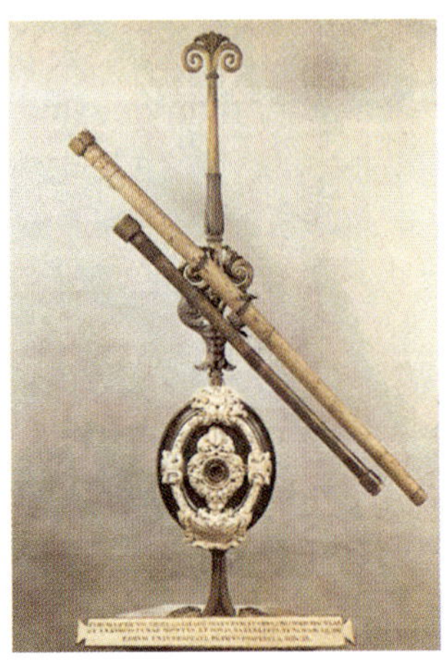

Photo Credit: Public Domian- from Singer, Charles, Studies in the History and Method of Science. Vol. ii Oxford: At the Clarendon Press, 1921.

- Federigo Giambelli constructs the first time bomb for use against Spanish forces besieging Antwerp, Belgium.
- The first water closet (toilet) is invented in England.
- Galileo begins constructing a series of telescopes, with which he observes the rotation of the sun and other phenomena supporting the Copernican heliocentric theory.
- Christian Huygens begins work on the design of a pendulum-driven clock.

- Robert Hooke develops the balance spring to power watches.
- Isaac Newton constructs the first reflecting telescope.
- The agriculture, mining, textile, and glassmaking industries expand.
- The concept of the scientific method of invention and inquiry is originated.
- Robert Boyle finds that gas pressure varies inversely with volume (Boyle's Law).
- Leibniz makes a calculating machine to multiply and divide.

1700 - 1800

A cross-section sketch of Bushnell's Turtle submarine. Image courtesy of Wikipedia

- The Leyden jar stores a large charge of electricity.
- The Industrial Revolution begins.
- James Watt makes the first rotary engine.
- Louis XV of France establishes the Ecole des Ponts et Chausses, the world's first civil engineering school.
- James Watt patents his first steam engine.
- Jesse Ramsden invents the first screw-cutting lathe, permitting the mass production of standardized screws.
- David Bushnell designs the first human-carrying submarine.
- John Wilkinson installs a steam engine to power machinery at his foundry, in Shropshire, the first factory use of the steam engine.
- Abraham Darby III constructs the world's first cast iron bridge over the Severn River near Coalbrookdale.
- Claude Jouffroy d'Abbans powers a steamboat upstream for the first time.
- Joseph-Michel and Jacques-Etienne Montgolfier construct the first passenger-carrying hot air balloon.
- Joseph Bramah designs his patent lock, which remains unpicked for 67 years.
- British civil engineer John Rennie completes the first building made entirely of cast iron.

1800 - 1825

- Automation is first used in France.
- The first railroad locomotive is unveiled.
- Robert Fulton begins the first regular steamboat service with the Clermont on the Hudson River in the United States.

ROBERT FULTON'S CLERMONT, THE WORLD'S FIRST SUCCESSFUL STEAMBOAT.

- Chemical symbols as they are used today are developed.
- The safety lamp for protecting miners from explosions is first used.
- The single wire telegraph line is developed.
- Photography is born.
- Electromagnetism is studied.
- The thermocouple is invented.
- Aluminum is prepared.
- Andre Ampere shows the effect of electric current in motors.

1825 - 1875

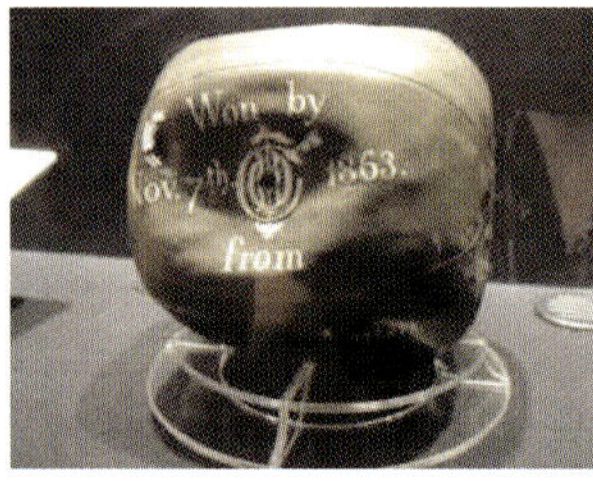

Information and picture obtained from the National Soccer Hall of Fame which is located in Oneonta, NY. Please visit their web site for additional information about the facility: http://www.soccerhall.org/
.

- Rubber is vulcanized by Charles Goodyear in the United States. In 1855, Charles Goodyear designed and built the first vulcanized rubber soccer balls.
- The first iron-hulled steamer powered by a screw propeller crosses the Atlantic.
- The rotary printing press comes into service.
- Reinforced concrete is used.
- Isaac Singer invents the sewing machine.
- Henry Bessemer originates the process to mass-produce steel cheaply.

- The first oil well is drilled near Titusville, Pennsylvania.
- The typewriter is perfected.

1875 - 1900

- The telephone is patented in the United States by Alexander Graham Bell.
- The phonograph is invented by Thomas Edison.
- The incandescent light bulb also is invented by Edison.
- The steam turbine appears.
- The gasoline engine is invented by Gottlieb Daimler.
- The automobile is introduced by Karl Benz.

1900 - 1925

Photo Credit: Al Gomez

- The Wright brothers complete the first sustained flight.
- Detroit becomes the center of the auto industry.
- Stainless steel is introduced in Germany.
- Tractors with diesel engines are produced by the Ford Motor Company.
- The first commercial airplane service between London and Paris commences.
- Diesel locomotives appear.

1925 - 1950

- Modern sound recordings are introduced.
- John Logie Baird invents a high-speed mechanical scanning system, which leads to the development of television.

Image Copyright Erard Bruno

- The Volkswagen Beetle (left) goes into production.
- The first nuclear bombs are used.
- The transistor is invented.

1950 - 1975

- Computers first enter the commercial market.
- Computers are in common use by 1960.
- The first artificial satellite Sputnik 1 (USSR) goes into space.
- Explorer I, the first U.S. satellite, follows.
- The laser is introduced.
- Manned space flight begins.
- The first communication satellite Telstar goes into space.
- Integrated circuits are introduced.
- The first manned moon landing occurs.

1975 - 1990

- Supersonic transport from the United States to Europe begins.
- Cosmonauts orbit the earth for a record 180 days.
- The Columbia space shuttle is launched and reused for space travel.
- The first artificial human heart is implanted.

Image courtesy of NASA

1990 - Today

- Hubble Space Telescope (HST), carried into orbit in 1990 in the space shuttle Discovery. Built from 1978-1990, the HST cost $1.5 billion.
- The Internet Society is chartered, and 1,000,000 host computers are connected in a network.
- Computer processor speed is dramatically improved.
- The Channel Tunnel (the Chunnel) between England and France is completed.
- The first rendezvous of a NASA spacecraft with the Russian Mir space station.
- MP3 Audio format traded widely though computer servers.
- World's tallest building opens in Kuala Lumpur, Malaysia (1,483 feet).
- Global Positioning Satellite (GPS) technology is declassified, resulting in hundreds of safety, weather and consumer applications.
- High-Definition Television signals/products available.
- The genetic code of a human chromosome is mapped.
- Robots walk on Mars.

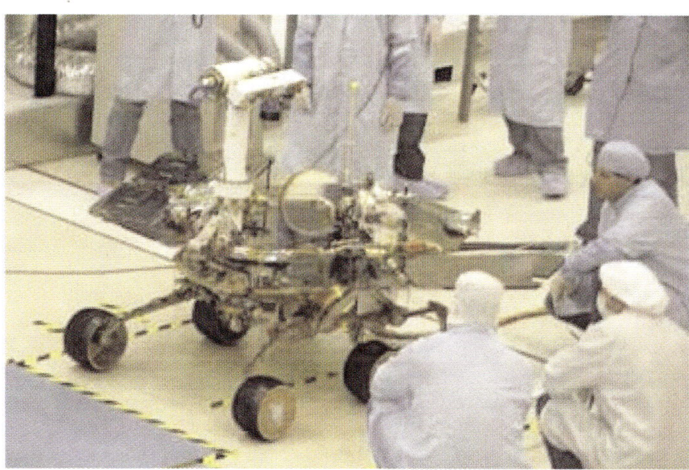

Image courtesy of NASA

The Great Pizza Factory

The history of Ford Motor Company is, in many ways, the quintessential story of the American dream. With big ideas and little cash, Henry Ford set out in 1903 to make the automobile accessible to every American. Almost one hundred years later, his company has become the world's biggest producer of trucks and the second biggest producer of cars and trucks combined.

In 1907, Henry Ford announced his goal for the Ford Motor Company: to create "a motor car for the great multitude." At that time, automobiles were expensive, custom-made machines. Ford's engineers took the first step towards this goal by designing the Model T, a simple, sturdy car with no factory options -- not even a choice of color. The Model T, first produced in 1908, kept the same design until the last one -- number 15,000,000 -- rolled off the assembly line in

Moving assembly line at Ford Motor Company's Michigan plant. Image courtesy of Ford Motor Company

1927. From the start, the Model T was less expensive than most other cars, but it was still not affordable for the "multitude." Ford realized he needed a more efficient way to produce cars in order to lower their cost. He and his team looked at other industries and found four principles that would further their goal: interchangeable parts, continuous flow, division of labor, and reducing wasted effort.

Your team will design an assembly line that can be adapted to handle variations in the desired pizza products. This activity will have several important parts. Teams will research assembly lines and the history of assembly lines. This part of the assignment could result in an oral presentation or written report. Teams will sketch preliminary designs for their assembly lines based on their research as well as the time required to make 1-5 sample pizzas. Teachers may time the students to see how close the actual time to make 1-5 pizzas compares to the time they originally proposed. Each team can complete peer evaluations regarding the quality of the pizzas made by other teams using the assembly line techniques.

In this activity, students work in teams to design an assembly line to make pizzas. This assembly line should be able to be modified to handle any type of pizza style (e.g., deluxe or just cheese and sausage). Students will need to locate local pizza supply companies and makers. They may want to visit food production companies to learn more about aspects of production, how resources are managed, and what steps are taken to ensure quality. Students will need to analyze costs and consider the best ways to customize the pizzas. Time and resources permitting, students may be able to carry out the tasks for a fundraiser or a team lunch.

RESOURCES

http://www.pbs.org/wgbh/aso/databank/entries/dt13as.html This site overviews Henry Ford's creation of the assembly line and the four main objectives.

http://www.ford.com/servlet/ecmcs/ford/index.jsp?SECTION=ourCompany&LEVEL2=heritage To learn more about the realization of an American dream and its expansion visit the web site for an interactive timeline, the Ford story, Henry Ford quotes and the "Spirit of Ford" video.

PROFILES OF HISTORICAL ENGINEERS

Over the course of history, many engineers have had great impact on society. Engineers design the bridges we cross, the roads we travel on, the automobiles we drive, and the computers we operate. They make up the backbone of society, improving the now and inventing the future. Without engineers, society would slowly progress as people thought of new ways of solving problems, whereas engineers spend their entire days solving problems. The important thing to realize about history is that it involves people. Much of it is based on biography. To understand the story of an invention, one should investigate the people involved because an invention and its inventor are always woven together. When studying the details of history, people find themselves more interested as they get to know more about the people involved and the challenges they were up against.

Engineering has been around in some form for thousands of years. We humans have always prided ourselves in making objects that will accomplish a task more efficiently that will allow people to do things they have never done, or that will allow them to go places they have never been before. Who were those elite people who paved the way for the rest? A brief list of histories greatest engineers follows.

a. Nolan K. Bushnell (1943-)

Nolan K. Bushnell is considered as the father of the video game industry. He founded Atari in 1972 and launched the video game revolution in that same year with Pong. He sold Atari in 1976 for 28 million dollars and founded "Chuck E. Cheese's" restaurant, selling that as well after a few years.

In an interview with Joyce Gemperlien of the San Jose Mercury News, he traced the spark that started his career to Mrs. Cooks' third grade class when he had an assignment on electricity. He put together his contraption and showed the entire class, then went home and found his flashlight, wires, and "old stuff" around the house and began to tinker with it, and he has never stopped.

Bushnell holds several patents on some of the basic technologies for many of the early video games he made and some he is working on. He is the inventor/co-inventor for many worldwide patents in various other fields.

b. Thomas Alva Edison (1847-1931)

Thomas Alva Edison is known to have said, "If we did all the things we are capable of doing, we would literally astound ourselves." Edison was a man of dream and determination. Applying his curiosity to the world around him, Edison managed to lay the groundwork for many technological advances still appreciated today.

In 1862, at the age of fifteen, Edison left his home in Port Huron, Michigan, to work as a telegrapher. He roamed the United States and Canada delivering telegraph messages. In 1869, he moved to New York where he worked on inventions related to the telegraph. He came out with the Universal Stock Printer, a device that would automatically print out stock telegraphs. For this novel invention, Edison received $40,000 in compensation, which in that day was enough money to enable Edison to become a full-time inventor.

He completed one of his greatest inventions in 1877. The cylinder phonograph was able to record and playback using tin foil and needles.

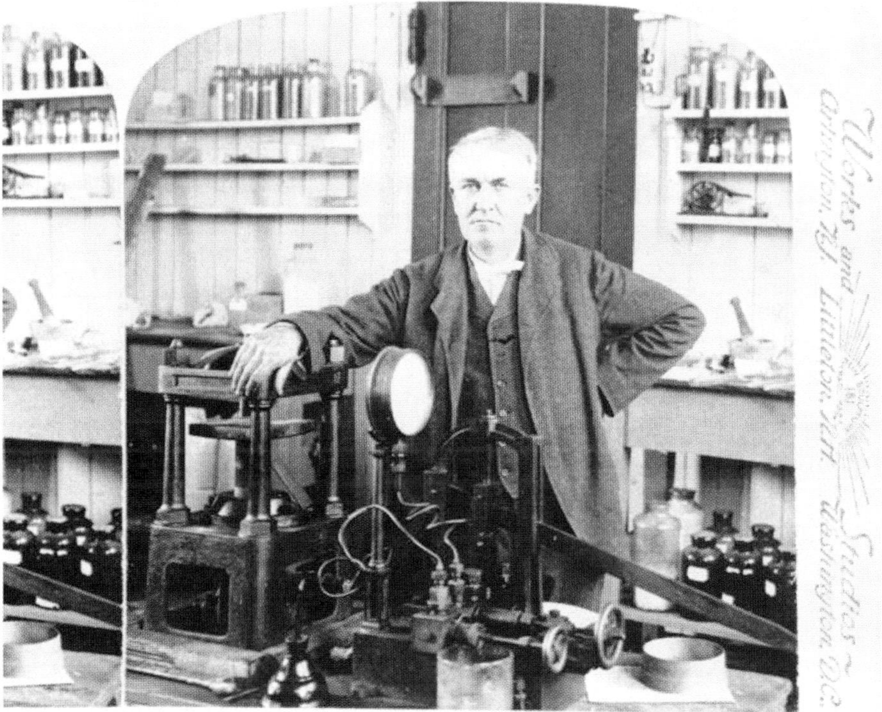

Thomas A. Edison in his laboratory in New Jersey, 1901. Credit: Underwood & Underwood, publishers. "The most famous inventor of the age—Thos. A. Edison in his laboratory, East Orange, N.J., U.S.A." 1901. Prints and Photographs Division, Library of Congress.

Edison took out a patent on his new machine on February 19, 1878, and to promote the phonograph, he sold the manufacturing and sales rights to a new company being formed called the Edison Speaking Phonograph Company. Edison received $10,000 plus 20% of the company's profits. Edison would return to the phonograph later in his career, but for many years he concentrated instead on his most famous work, the incandescent light bulb.

The concept of electrical lightning was not new in 1879 when Edison entered the field. However, no one had invented a product that would be practical for home use. Edison was challenged to create an electrical source of light and had to make the bulbs durable, long lasting, safe, and economical. Edison did produce a light bulb that met all of these requirements, and it only took him one year.

Over his 84-year life span, Edison took out 1,093 patents and started ten different companies. One of these companies became General Electric, a leader in the electronics industry today. Edison was a man of innovation who summed up his life's work by saying "Genius is one percent inspiration, ninety-nine percent perspiration."

c. Leonardo da Vinci (1452-1519)

Leonardo da Vinci had an amazing ability to imagine mechanical inventions that would one day become commonly used, especially in the field of warfare (helicopters, tanks, artillery). Da Vinci's handicap was his inability to read Latin, which prevented him from learning from the common scientific writings of the day which focused on the works of Aristotle and other Greeks and their relation to the Bible. Instead, da Vinci was forced to develop his ideas solely from his observations of the world around him. He was not interested in the thoughts of the ancients; he simply wanted to use his engineering skills to improve his environment.

He did use his genius in a wide range of subjects, but little of his work had any relationship to that of his generation. He designed, built, and tested, but essential aspects of development were always lacking from his innovations. He could do many great things, but he couldn't do them all. He lacked a community of engineers to work and think along with him. Being so far ahead of his time related to the development of engineering, it was centuries before many of his innovations came into existence.

Da Vinci's life was completely focused on thoughtful creation. He painted, he tried sculpture, he was an architect, and he engineered hundreds of useful devices and many others that never came to life. He was a respected member of the community and was consistently in the employment of people of great wealth.

To obtain his first paid work as an engineer he wrote a letter claiming that he could do the following: construct movable bridges; remove water from the moats of castles under attack; destroy any fort not built of stone; make bombs, dart throwers, flame throwers, and cannon capable of firing stones and making smoke; design ships and weapons for war at sea; dig tunnels without making any noise; make armored wagons to break up the enemy in advance of the foot soldiers; design buildings; sculpt; and paint.

With that résumé, he was hired with the title of Painter and Engineer to the Duke. Did he do all the things that he said he could? We don't know, but we do know that, among other things, he worked diligently perfecting the canal system around Milan, Italy.

Da Vinci's chief job, during the period he was employed by the Duke was to produce spectacular shows for the entertainment of the wealthy people who came to court. He produced musical events, designed floats, and delighted the court with parades that dazzled the eyes and with flying devices that amazed the audience. The interesting thing is that during this time he wrote over 5,000 pages of notes, detailing every imaginable kind of invention. This truly was a great engineer at work.

d. Isambard Kingdom Brunel (1806-1859)

Isambard Kingdom Brunel (1806-1859) was quite possibly one of the most diverse and most capable engineers in history. Brunel's projects showed his diversity and excellence. Brunel designed and built various projects ranging from railways to bridges to ships.

He was the son of engineer Marc Isambard Brunel, and returned to work his father's office at the age of 16 after his father sent him off to college at the age of 14. His first position was as an apprentice on the early stages of the Thames tunnel. He advanced to resident engineer in charge of the project in 1826 at the age of 20. On May 18, 1827, the Thames River broke through the tunnel his father designed. Isambard Kindom dove down into the river in a diving bell to inspect the damage that would end up being repaired with bags of clay and gravel. Eight months later, on January 12, 1828, the river broke through a second time and Isambard was seriously injured during rescue operations but did manage to save the lives of several men. All work in the tunnel stopped for seven years because of financial problems.

Brunel was the lead designer for the Great Western Railway project. He designed the track layout and the track itself for the project. Included in his designs were the ocean liner that would take the passengers from the end of the train track to America, and even the lamp posts in the train stations. The Great Western Railway was, at the time, the longest railway ever conceived.

e. Charles Proteus Steinmetz (1865-1923)

"No man really becomes a fool until he stops asking questions," said Charles Steinmetz. Steinmetz was a pioneer in the field of electrical engineering. Without his passion for understanding, electricity wouldn't exist in the same way it does today.

One of his first research jobs at General Electric was on hysteresis, which is the loss of power due to magnetic resistance. His research led him directly to alternating current, which solved the power loss issue but not the complete problem. No theories existed at the time for alternating current. Over the next 20 years, he successfully developed these theories. Without his theories, many inventions using electric power and motors would never have been created later.

An incident that occurred after Steinmetz retired illustrates how valuable he was as an engineer. One of the generators he had worked on was broken and none of the technicians had any idea on how to fix it. They asked him to come back and fix the problem. With chalk he marked the malfunctioning part with an "X." He gave them a bill for one thousand dollars and they asked him for an itemized invoice. He sent back an invoice, which said: "Marking chalk "X" on side of generator: $1. Knowing where to mark chalk "X": $999."

Steinmetz took out 200 electrical patents in his lifetime and considered his work in the field of electromagnetism to be his greatest accomplishment. He invented the metallic electrode arc lamp and worked to develop devices to protect power lines from lightning bolt strikes. Steinmetz even developed designs for electric vehicles, but the production of these new cars and trucks was stopped when he died.

f. Henry Ford (1863-1947)

Henry Ford was born on July 30, 1863, in Wayne County, Michigan. As a young man, Ford disliked school and farm life, so he walked to Detroit in search of a job. He found just what he was looking for in a small machine shop working as an apprentice. In this shop, Ford learned of engines and how they work. He worked for several years learning everything he could about engines, then opened his own shop back in his home town of Dearborn, MI.

Ford worked his way up to chief engineer at the Detroit Edison Company, and this is where he designed his first vehicle. It was called the Quadricycle and had a buggy frame with four bicycle tires as wheels. Ford continually revised his designs for vehicles until he founded the Ford Motor Company in 1903. The Model T was introduced in 1908 and stayed on the market for 19 years. It was due to Henry Ford's advancements in production technology that he was able to market the Model T so well.

Prints and Photographs Division, Library of Congress.

"His general policy of resistance to labor is as out of step with the times as the oxcart."
The Washington News, in a 1937 editorial against Ford

The creation of the assembly line in 1913 allowed the production of automobiles in greater quantities than previously possible and at lower cost. These reduced costs allowed the average American to own an automobile. Along with the assembly line, Ford instituted a minimum wage at his plant of $5.00 a day claiming that it increased productivity. One little known fact about Henry Ford and his contributions to society is the five-day workweek. Up until Ford created his company, most businesses operated on a six-day workweek. Henry believed that the extra day off would let people have more leisure time and make them more productive at work. Society has operated on the five-day workweek ever since.

g. Dr. Robert Goddard (1882-1945)

Dr. Robert Goddard grew up in New England and invented the process used for liquid-fueled rockets. It was his design that first allowed a rocket not fueled by solid fuels like black powder to get off the ground. Designs similar to his are used in the rockets that launch astronauts into space, deliver warheads to distant countries, and carry the communications and espionage satellites into various orbits around the Earth.

Dr. Robert Goddard, Image Courtesy of NASA

His system design enabled man to travel to and walk on the moon and allowed one to launch a multi-ton warhead to reach the other side of the world.

g. Wilbur Wright (1867-1912) and Orville Wright (1871-1948)

Growing up in a rather small area of the country at the time, Orville and his brother Wilbur enjoyed tinkering around in a workshop making pointless little mechanical objects. The brothers were still young when they decided that they wanted to become members of the new field of aeronautical engineering. So, both brothers taught themselves how to read and write and began to keep up with world engineering issues by reading the paper and studying books. They were most curious, however, about the work of Otto Lilienthal (a German scientist who tested gliders and kites). Wilbur and Orville were determined to design the first powered flying model. While they were working on their newest project of testing curved wing design, Orville built the first machine that we call today a wind tunnel. Using this machine, the Wright brothers could build scale model wing designs and test in the wind tunnels the levels (quantitative) of lift and drag on any shape they wanted. This major step in history set the Wrights ahead of every other leading scientist in the world. Here is an excerpt from the Journal of the Western Society of Engineers, December 1901:

The Wright brothers found a place along the Atlantic coast where the wind blew rather consistently at 16 to 25 miles per hour. Then they tied their early plane to a rope so the wind would lift the plane in one place over the ground much like a kite. This allowed for easy testing of their plane designs.

By and far, the tasks of piloting a successful flight is what the Wright brothers are credited for. However, the way in which they did these experiments should be considered their greatest achievement. Without the practices they developed and the extensive testing they did, the flight never would have happened.

h. Grace Murray Hopper (1906-1992)

Official U.S. Navy Photograph, from the collections of the Naval Historical Center.

Grace Murray Hopper was a computer engineer and Rear Admiral in the U.S. Navy. She developed the first computer compiler in 1952, and the computer program language COBOL. Grace popularized the computer program term "bug" when she discovered a moth jammed in an early computer. In 1983, by special presidential appointment, Hopper was promoted to the rank of Commodore. Two years later, she became one of the first women to be promoted to the rank of Rear Admiral. In 1986, after forty-three years of service, Rear Admiral Grace Hopper retired. At 80 years old, she was the oldest active duty officer at that time. She spent the remainder of her life as a senior consultant to Digital Equipment Corporation.

Grace Hopper received many honors over the course of her lifetime. In 1969, the Data Processing Management Association awarded her the first Computer Science Man-of-the-Year Award. She was the first person from the United States as well as the first woman to be made a Distinguished Fellow of the British Computer Society, in 1973. In September 1991, she was awarded the National Medal of Technology, the nation's highest honor in engineering and technology.

i. Joseph B. Strauss (1870-1938)

Joseph Strauss "Photo courtesy of the Golden Gate Bridge, San Francisco, CA, www.goldengatebridge.org."

The idea of connecting each side of the Golden Gate strait near San Francisco with a bridge was proposed as early as 1872 by Charles Croker. As late as 1916 most engineers said it would be impossible. However, Joseph Strauss said it could be done, and for between $25-$30 million. He did have experience in such matters as the designer of nearly 400 bridges. Strauss dedicated himself to the bridge, informing political leaders that the bridge could pay for itself with tolls alone.

The War Department had the final say in approving construction because of the impact of the structure on shipping and military logistics, and it owned land on both sides of the Golden Gate strait. A provisional permit was given on December 20, 1924. However, Strauss' work was not complete. This bridge was not popular with owners of ferry boat companies and others. However, the bridge project prevailed, and construction began on January 5, 1933.

The Golden Gate Bridge has a span of 1,280 meters, and connects San Francisco to Marin County. When this bridge was opened, it held the world record for its extensive length. Since the bridge was completed, millions of cars have passed over it.

The Golden Gate is linked from one end to the other with long cables that are suspended through the air. The total length of these cables is long enough to circle the entire world three times. It took two unbelievably long years to hang these cables back and forth creating the design.

The total weight of the bridge, including bridge, anchorages, and approaches after the re-decking project of 1986 is 887,000 tons or 804,700,000 kilograms.

j. Leo Szilard (1898-1964)

After studying engineering in Budapest and Berlin, Szilard moved to London in order to escape Nazi persecution. In 1933, he obtained a patent for the nuclear chain reactor. He first tried to use the elements indium and beryllium to create a chain reaction but discovered they would not work. In 1939, two years after moving to New York City, he learned about fission and subsequently discovered uranium would work for such a reaction. Worried about the Germans developing an atomic bomb, Leo recommended the U.S. government immediately start work on the atomic bomb. As a result of his letter to the U.S. government, he convinced them to launch the Manhattan Project. Finally, on December 2, 1942, Szilard and his partner Enrico Fermi successfully generated the first controlled nuclear chain reaction (explosion).

Even though he is credited with building the first atomic bomb, he advocated against its use. He organized an opposition to the May Johnson bill, which would have given the responsibility of all nuclear energy to the U.S. military. In 1955, Szilard and Fermi were given a patent for the nuclear fission reactor. Szilard helped the field of nuclear engineering and made contributions to political science, nuclear physics, statistical mechanics, genetics, and to molecular biology.

Szilard was a collaborator with Albert Einstein on 45 patent applications over seven years in six countries, mostly on refrigeration technology.

Your greatest engineer report

Thousands of engineers have influenced our lives. During their careers they may have worked on hundred of projects or perhaps only one that was updated year to year. Your assignment is to write a two-page essay about the person you think is the greatest engineer of all time. Your essay should not include any of the engineers listed in this textbook. A brief outline will help you lay out your report.

Use a separate sheet of paper to create your outline. When you are finished writing your outline, check it over with your instructor to make sure it is in the correct format and has the correct information in it. After your instructor approves the outline, write at least three sentences for each object listed in your outline. Use the program PERRLA within Microsoft Word to ensure that your paper is written in the correct format.

Submit your paper to ag@glpbooks.com and it may appear in the next version of the textbook.

The “A” essay achieves all the goals of "C" and "B" essays, plus it relates the issues and arguments to your own personal experience. It states your views on the issues and how they apply in your own life.

The “B” essay achieves all the goals of the "C" essay, plus it compares and contrasts the positions of the authors. It expands and extends the authors' ideas beyond what is explicitly stated in the readings.

The “C” essay demonstrates you did the reading, understand the issues involved and grasp the authors' positions on those issues. It explains the supporting reasons and arguments for the positions on both sides of each issue. Therefore, it explains both what the authors believe (theirpositions) and why they believe it (their reasons and arguments).

The content of your essay is more important than the style of your-writing. But you should be aware that content and writing technique are closely linked. You may know* the material, but if you cannot convince the reader that you know, your grades will disappoint you. There are four general standards which must all be observed:

1. Your writing must be clear: Be sure to say exactly what you mean. It is not sufficient to hint or suggest your meaning. You must state your points explicitly so there is no doubt about your meaning. Students often ask, "Couldn't you figure out what I meant?" It isn't the reader's job to guess your meaning. It is your job to say it clearly. Even when I suspect that a student knows an answer, if it is not clearly stated, I will not give credit for what is not said.

2. Your writing must be unambiguous: Although this is closely related to clarity, it is so important that it deserves separate mention. Your writing should not be open to multiple interpretations. Statements that are too general can cover too much ground. Poor grammar or poor word choice can confuse meaning. You must communicate your ideas so there is no doubt about your meaning.

3. Your answers must be complete: Partial answers deserve only-partial credit. To get full credit, you must answer the entire-question, not just a part of it, and certainly not some other question (like the one you studied for). Multiple-part questions require multiple-part answers. Giving a complete answer to the specific question asked demonstrates your mastery of the material.

4. Your answers must be accurate: Being clear, complete, and unambiguous doesn't count for much unless you are also accurate. Silly mistakes or oversights can rob essays of their accuracy. (For example, writing, "Smith would agree with Jones.", instead of, "Smith would disagree with Jones.") Unless you reread your essay for accuracy, you run the risk of letting little mistakes rob your writing of its intended meaning. Take the time to review your work for accuracy.

* Passive Understanding vs. Active Mastery: Students sometimes confuse passive understanding with active mastery. Because material makes sense (passive understanding) when they read it, or when it is discussed in class, they think they "know" it and are disappointed when they earn a “C”. Active knowledge and mastery require not just that you understand the material when someone else speaks or writes about it; they require that you, yourself, are able to clearly and accurately explain what the material means and what it implies. Just as passive understanding of a word does not guarantee that you can use it correctly, passive understanding of a subject is not the same as knowing it. Passive understanding earns a “C”, at best. Active knowledge earns a “B”. Mastery earns an “A”.

Resources

www.perrla.com APA PERRLA correctly creates references and citations that conform to APA format. APA PERRLA takes you through a simple, step-by-step PERRLA Wizard process

Special thanks to Professor Warren Z. Weinstein of the Philosophy and Asian Studies Departments at California State University, Long Beach for the grading rubric used in this activity.

Chapter 3

Expectations of an Engineer

EXPECTATIONS

What are our parents, teachers, and our employers really looking for? After careful study, this section will define some of the important "soft skills" or "interpersonal skills" non-technically-oriented characteristics people and employers desire. These skills are usually common sense, but don't be fooled into thinking they are unimportant.

SKILLS

One of the hardest parts of getting a job is to retain the learned skills to achieve the position. Professionals make every effort to keep up their skills through professional development, courses, seminars, etc., that provide updates to their collection of skills. People have trouble learning these interpersonal or soft skills out of a book. Learn to develop these skills in various areas of your life

The Wisconsin Governor's Work-Based Learning Board has defined several skills needed for success in the workplace. A good student or employee does the following:

- Communicates clearly with supervisor and others
- Acts professionally
- Learns effectively
- Manages self responsibly
- Plans for change (flexibility and adaptability)
- Plans for personal and professional growth
- Works productively
- Recognizes safe and unsafe work habits
- Demonstrates proper safety procedures
- Maintains a safe and healthy work environment

BE ON TIME

It is very important to be on time and to stay on schedule.This critical workplace issue is often overlooked. Your responsibility is to be on time and having the materials needed to get the job done.

COMMUNICATION

A successful person will develop the ability to communicate well with others. It has been said that sometimes more work can get done in eighteen holes of golf than in a week of meetings. The point is that people need the ability to talk effectively to each other, no matter what the setting. Competitors are doing everything they can to get people's business, and if a personal relationship is established with a customer, it can strengthen a company's position. The ability to negotiate, to effectively make your case, to make clear presentations, to use appropriate body language, and to understand cultural differences are all valuable to a successful engineer.

THINK "OUTSIDE THE BOX"

An engineer must be able to think logically, of course, but an engineer must also learn to be creative, and innovative, developing new means of looking for solutions to problems. We sometimes call this "thinking outside the box"- going beyond the boundaries of what already exists (the box). What if that box is a "sphere"? Each situation will present its own form, and engineers must determine how to solve the problem. Here is a question for evaluation: A group of experienced engineers and fourth graders sit down at a table together and are given the same problem to solve. Who will have the most creative solution? Answer: The fourth graders are more likely to have the more creative solution because they have not yet been trained to think only certain ways about problems. Seasoned engineers have gone through schooling and formal training and have learned "the best way" or "the usual way" to think about solving problems. Their thinking is often limited to the ideas "inside a box", thereby limiting their imagination.

POSITIVE ATTITUDE

Like many professionals, engineers must have a positive attitude, a love for their job, and an ability to work along side others. Too often, people get complacent in their jobs and "get into a rut". Becoming excited about projects, becoming involved with the details, and having a willingness to learn are positive traits worth having.

EDUCATED RISKS

John Young on the moon's surface. Image courtesy of NASA

Be comfortable asking, "What happens if? How would this work?" Take educated risks that make sense, not risks that are long shots or that might have very bad consequences. When you work in teams, your teammates may think of ideas you wouldn't have thought of, and they may help you decide if your idea is worth trying.

BUSINESS AND FINANCE

Many companies want their employees to have a solid understanding of business and finance as well as the ability to explain what they are doing throughout a project to people that are not necessarily engineers. Engineers might be involved in proving to a company's management that a project if worth pursuing financially.

LIFELONG LEARNING

Lifelong learning is critical for people who want to excel in their field, and engineering is no different. As standards, practices, and the wants, and needs of society change and evolve, so must the engineer and company. If the engineers in a company do not stay up to date, it can jurt a company's long-term success.

CONFLICT RESOLUTION

The possibility that members will disagree always exist. Everyone comes from different backgrounds, values, training, and other variables that give the team significant advantages versus a team that is assembled exclusively of similar people. Conflict resolution can help you navigate those tough times when group members have issues with each other. Everyone must consider the interests and suggestions of other people while working on a team. Each person needs to be able to step back from the situation to evaluate it positively. Many employers mention this trait as one of the most important they are looking for in an employee, particularly when that employee will be working with a team of people.

It takes cooperation from everyone on a team to achieve a goal. Photo Credit: Al Gomez

WORK WELL WITH OTHERS

Employers state that the ability to cooperate with others is an essential trait they want in an employee. People who work with each other need not be best friends, but they might develop strong relationships. If people have friendships at work, they tend to enjoy the time they spend there more. This creates a positive environment that increases productivity. Co-workers who do not get along with each other for personal reasons still must determine how to work well together no matter their feelings.

CRITICISM

Friends on a golf course often criticize each other, but only for fun. Photo Credit: Al Gomez

Employees must be able to take criticism, which is a statement about how something may be done better. You do not have to like what you hear, but you must be receptive to others comments, evaluate the comments made, and make a decision on whether to alter the area of question or not. Constructive criticism can provide essential feedback and help you improve your performance.

Build your sample resume

Résumés are important in a job interview. These documents provide an employer with an account of your education and work experience. Often, they also list your special skills, hobbies, and/or interests.

A résumé is a one-page snapshot of you. The résumé typically should be no more than one page. In this activity, you will use a template in a word processing software program to insert information about yourself as if you were applying for a job. You may not have a lot of information to work with at this stage in your life, so go ahead and make up some of the details as practice.

There is a wide variety of appropriate formats for résumés, so ultimately you have to decide upon a format that you will be happy with. Regardless of the specific format, every résumé must present your name, address and phone number; your educational background; your previous employment (if any); and extracurricular activities. The following simple format is merely one of many. Key elements are highlighted in bold type.

Make sure that you get as many people as you can to read and critique your résumé. Listen to and evaluate all the comments you get, and then customize the résumé to fit your needs.

Your instructor will assign you the template and number of jobs, hobbies, and skills to present in your résumé.

Résumé Checklist

Information

- Objective is clearly stated.
- Clearly convey how your education, experience, activities, and honors support the objective.
- Experience, education, and/or skills segments are effective.
- All activities, honors, and other data are appropriate for the employment and the reader.

Organization

- Name and key headings stand out.
- Information within each heading is ordered from most to least important.
- Experience segment is arranged to highlight your strengths and career objective.

Style

- Language is simple, direct, and precise.
- Noun phrases are consistently used for headings.
- Strong verb phrases or clauses are consistently used to describe experience, skills, and activities.
- Parallel structure is used effectively.
- Have no errors in grammar, punctuation, or spelling (no typos whatsoever).

Important things to remember about résumés

- A résumé should be crafted for a specific position or job. This shows the employer that you care about the position. Your résumé's objective section should reflect some specificity as a result.
- You may need to change your résumé for "scannable format," if you apply for jobs with larger engineering companies. These companies use a computer to scan each résumé submitted to them for keywords. If so, you should consult your career services office or look at the company's recruitment information to determine these keywords and use them on your résumé. The scanner will target you and your résumé as a potential employee more quickly.
- If possible, keep your résumé to one page. Many employers frown upon résumés longer than one page.
- The format of a résumé is important. Your information should be arranged to be visually pleasing and readable. Use the following strategies in terms of format:
  - List dates in reverse chronological order (starting with your most recent information first)
  - Use at least 11-point font size (preferably 12-point) and typeface such as Times New Roman
  - Use one-inch margins around the page

- Headings such as Education, Work History, Skills, Honors and Activities, etc., should be in bold, can be in capital letters, and should be the only information on that line of the page
- Your résumé should be printed using a high-quality laser printer on excellent bond paper (at least 40-pound ivory-colored or off-white paper)

Here is a list of recommended action verbs you may want to use, categorized by topic:

Human Relations

worked with	taught	helped
volunteered	served	assisted
interacted	counseled	trained
sponsored	directed	guided

Research and Design

researched	experimented	observed
analyzed	tested	assessed
solved	verified	designed
discovered	devised	created
investigated	evaluated	

Communications

drafted	edited	addressed
designed	revised	interpreted
composed	prepared	lectured
published	taught	conducted
presented	instructed	published

Management

managed	contracted	saved
rated	maintained	scheduled
evaluated	established	sold
devised	negotiated	verified
planned	controlled	produced
organized	purchased	improved
coordinated		

(See actual Sample Résumé on next page.)

Sample Résumé Format

Your Name (centered)

College Address	**Permanent Address**
College Phone Number	**Permanent Phone Number**

Objective:	An entry level position in (state field or position sought).
Education:	• your degrees • your school • graduation date (or anticipated date) • grade point average and basis (e.g., 3.4/4.0)
Work Experience:	• your job titles (list most recent first) • the companies for which you worked • dates for which you worked for those companies • responsibilities in those positions
Activities:	• meaningful activities in which you have participated

Sample Actual Résumé

Jason D. Smith

222 Landon Hall Michigan State University East Lansing, MI 48824 (517) 555-1212 smith@pilot.msu.edu	37 Dilgren Avenue Port Anglican, VT 12225 (234) 555-1212

Objective:	An internship in the manufacturing area of a mid-sized company
Education:	Bachelor of Science in Mechanical Engineering Michigan State University, East Lansing, Michigan Expected Date of Graduation: May 2003; GPA: 3.6/4.0
Computer Skills:	MS Office 2000, Lotus, C++, MATLAB
Employment:	May 2002—August 2002; Production Engineer Labardee and Sons, Aberdeen, Texas • Developed new production plans for the Labardee grinding machine • Consulted with primary Labardee clientele June 2001—August 2001; Machinist Rawlins Machining, Deer Park, New York • Produced updated schematics for all standing jobs • Instituted changes in assembly line procedures
Honors/Activities:	Ames Scholarship, Dean's List (4 semesters), MSU Swim Team, racquetball
References:	Available upon request

TENDENCIES

This fisherman's smile and focus on the big fish he just caught lets us know he is very interested in what he is doing. Photo Credit: Dennis Skurulsky

A team member should possess the ability to sense the tendencies for other team members. This is attempt to develop the ability to "read" unspoken clues in other team members – their hidden signals and body language. Posture says a lot about how interested people are in a given situation. If the shoulders are loose and forward position and the person is leaning to one side, they may be uninterested in the topic at hand, not ready to answer a question or engage in a conversation. However, if someone is making good eye-contact, and appears attentive, they are likely in tune to what is being discussed.

Another good way to show you are interested in what people have to say is to lean forward slightly when they are speaking. We do this almost instinctively when we are truly interested and may fail to do this when we are disinterested or distracted. If you have problems looking directly into people's eyes, pick a point on their forehead not too far above the eyes and concentrate on that spot. Because it is so close to the people's eyes, they will not realize you are not looking directly at them. Some studies indicate that if you look at the other persons left eye – the one on the right as you face them – you convey the greatest sense of connection and concern.

So, be positive, lean a bit forward, and look almost into their eyes when someone is speaking in a small group, even if you are only moderately interested since body language is always important. A simple smile can take the edge off of a conversation and comfort a listener or speaker in a conversation.

SARCASM

Webster's defines sarcasm as a keen, reproachable expression or a satirical remark uttered with some degree of scorn or contempt. Sarcasm should never be used in a business or professional setting where people do not know each other well. Even if the group discussion has only one unfamiliar person, be professional. Sarcasm may have a place among friends who are amusing each other, but should be avoided in more formal situations.

FLEXIBILITY AND OPENNESS

Flexibility and openness are important qualities when dealing with people of different cultures. Each culture has its own idea of what things are acceptable in business situations. A person must be able to adapt and improvise if necessary, which includes respecting another culture and its customs. Staying open to new ideas and alternative perspectives may lead you to solutions and ideas you may not have developed on your own.

WRITTEN AND ORAL COMMUNICATION SKILLS

Written and oral communication skills are critical for success when studying or practicing engineering. Most employers require their employees to document their project and present their reports to others. This may take the form of an e-mail to a supervisor, a technical memo, a presentation to a group of engineers and technicians, or even a formal presentation to a group of directors or executives. Submitting weekly activity reports and other documents may be part of a job requirement. When writing these reports be sure to use correct grammar, complete sentences, and correct spelling. Spell check and proofread your work carefully. When speaking do not use slang or unnecessary jargon.

NETWORKING

Networking is a term that describes the ability to find opportunities and or meet new contacts by working through "chains" or "webs" of people you know. This is a great help in meeting your short-term and long-term career objectives. Networking can be described as getting together with people who are able to provide advice, lend support with projects, and make valuable connections for you. One contact can lead to another and then to another.

SELF-DISCIPLINE

To be successful in any challenging field, you must develop self-discipline. As you progress through school and on into your profession, you will need more and more self-discipline. You will need to develop good time management, punctuality, and preparation and organizational skills. As you acquire and develop these skills, you will find that you can apply them to many different situations, which in turn makes you more productive and successful.

SOCIAL INTERACTION

In your career, you likely will need to interact with a variety of people. These may include other engineers, technicians, scientists, production workers, labor union representatives, clerical staff, managers, directors, and executives. Your involvement in various social activities and organizations can be good preparation for future professional interactions. Employers look for people who have activities in their lives outside work. On a résumé or in an interview, include things like hobbies, community involvement, church activities, and volunteer positions.

Student Learning Activity

Your first job interview

Section 1: The interview

Interviewing can be a stressful experience since you typically are interacting with someone you don't know, trying hard to make a good impression. Here are a few points that can help in an interview:

- Have confidence.
- Be friendly.
- Speak to everyone in the room. Don't speak too softly or too quickly.
- Lean forward a bit toward the interviewer (body language suggests that you are interested in the conversation) and make good eye contact.
- Wear appropriate clothing. Business dress is always suggested. It's safer to risk being over-dressed.

As you interview, bring up past experiences that show that you have interest in the field for which you are interviewing. Discuss your qualifications. Bring a portfolio along so that if asked you can show examples of past work.

Section 2: Marketing your skills

During a job interview, show and tell about yourself. Use positive, reinforcing statements to show you are confident but not arrogant. Below is a list of helpful words to use when interviewing.

ASSIGNMENT A:

With a partner, quiz each other on these terms and their meaning. Consult a dictionary if you need help.

reliable
trustworthy
personable
punctual
loyal
flexible
productive
enthusiastic
compassionate
patient
organized
ambitious
hardworking
responsible
self-disciplined

ASSIGNMENT B:

Visit this website from the Western State College of Colorado: http://www.western.edu/career/Interview_virtual/Virtual_interview.htm. See if you can answer the questions correctly. For questions you answer incorrectly, think about why your answer was wrong.

Section 3: Interview questions

Interview questions are sometimes hard to answer, but if you practice answering them in advance, you will find them easier to answer. Think of three questions that you would want to ask someone you wanted to hire. Write these three questions on a piece of paper and turn them in to your teacher. Your teacher will go over the questions that the class has submitted and create a list of the best questions.

Break up into groups of two (not the same partner as before). You and your partner should imagine a job you want to interview for so you have specific examples you can discuss. Ask each other the list of questions and write down your partner's answers on a piece of paper, arranged as shown in your workbook.

Chapter 4

Systems and Optimization

SYSTEMS AND OPTIMIZATION

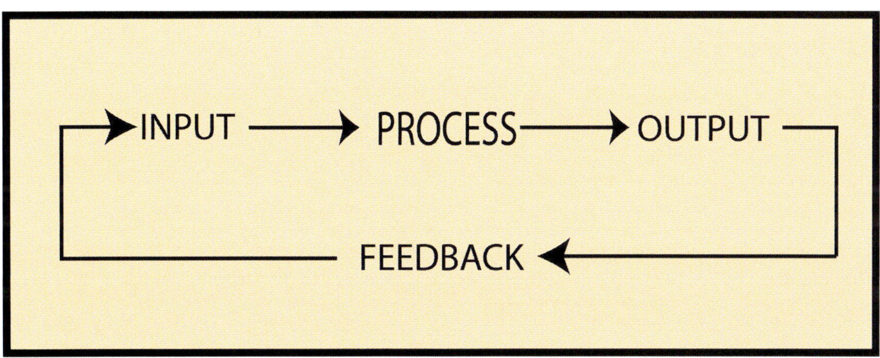

A system is a mechanism for achieving a desired result. It involves Input, Process, Output, and Feedback. Breaking problems or situations into these elements can simplify them for you.

For example, an automobile has systems and subsystems: mechanical systems (gears and pulleys), electrical systems (battery, wiring, and computers), and fluid systems. As the driver, you are the input part of the system as you feed the car information in the form of pressure on gas and brake pedals. You manipulate a steering wheel mounted to a column that controls the direction of the vehicle. The automobile converts the input to a process, obeying your signals and converting it into output.

If one of these systems fails, the driver wants feedback. In a vehicle, the gauges and displays give one form of feedback, from the fuel gauge to the warning lights that tell you something is wrong with your engine.

I. INPUT

Inputs can be varied but fall into several categories. They can be in many forms, however we will discuss the top three. These are people, information, and time.

People have the ability to control their own environments and do so through input. They can design products and systems in order to satisfy a want or need of a single person, or millions of people. People can make decisions without having data or information. These decisions are not always the best for the desired goal. Through ethical practices and sound decision making, people are always important in the systems process.

Data is everywhere in our lives, from the statement that a customer receives about their bank account to the box scores in the Sunday paper, data is all around us. Data and information are one of the most critical inputs that an engineer or designer can have in order to attack a problem.

For example, if you were to predict how many cars traveled on a road in a week's time, data would need to be collected. Civil engineers place counter boxes that have hoses attached to them that count the number of times vehicles pass over them every day. These boxes are placed in strategic places along travel routs to get the information and data that the engineer needs. This data can then be transferred and organized to tell a story. Was there a major business convention in town that week? Did the local school only have half days that week?

Time could be a characteristic that fits well within information, however it is significant enough to be put into its own category. Time is always a factor in the process. If an engineer knows that time is limited, they may make decisions that limits what can and cannot be done. Time can be an enemy and it can be a friend.

For example, if an engineer knew that they had three months in the summer to build a new roadway they could plan out what needed to be done each day and each hour. If the weather cooperated all summer, all the materials and machinery was available and no major setbacks occurred, the engineer could actually finish the project early.

II. PROCESSES

All systems have actions that are associated with them. These steps can be completed in many different ways and make up what are the processes in the systems loop. There are many different processes that can take place in this block of the loop including problem solving and production.

Typically, much of today's research, design and development is still being done in central locations, though actual production may be occurring in many different parts of the world. Engineers may find that their work will involve communication and input from colleagues around the globe. The typical foreign-based manufacturing facility is usually staffed by foreign national employees. Almost all of the production staff and skilled and unskilled labor are hired from the particular region. The corporate staff is usually involved in the initial start-up of the facility, the hiring and training of the employees, and the monitoring of the manufacturing process. Often, it is most efficient for a corporation to use local engineering talent, so many firms will recruit international students attending U.S. colleges and universities to fill these roles.

A similar situation exists for foreign firms that have established manufacturing facilities in the United States. Most of the on-site corporate staff will be from the home country, but the objective is to try to hire and train as many U.S. employees as possible. Foreign firms will invest time and money to train these individuals to adapt to their particular style of doing business.

Many of the ideas and concepts which have revolutionized the manufacturing industry have been a result of the globalization of world production processes. Engineers and managers are adopting some of the best manufacturing technologies which have been developed in many different countries. Today's world of manufacturing is often a blend of the best practices and procedures gained from the integration of systems and concepts from engineers around the world. Manufacturing is truly a global enterprise in the life of today's engineer.

IDEAS Sanitation Systems

CASE STUDY

The origins of the first sanitation systems were seen in ancient India, Babylon, Egypt, Crete, and Greece, going back as far as 1500 to 2500 BC. The citizens of the ancient Greece and Rome suffered greatly from refuse and sewage problems. Even with today's modern sanitation and waste disposal technologies long-term waste and trash disposal solutions are quite challenging due to population growth.

The modern era of waste management in the United States is only a little more than a hundred years old. Within this relatively short period, modern waste treatment systems have been developed and implemented. These developments have greatly enhanced disease control and health for most Americans.

The wastewater treatment systems in every city represent major engineering accomplishments. Although most people never give it much thought, the sewer systems under their streets required careful design, layout, and construction. This system of piping allows for an efficient and safe means of transporting waste and dirty water from throughout a city to a centralized treatment plant where it can be broken down and disposed of properly. A close look at your local sewer system will provide an excellent example of the contribution of Civil Engineering to society's health and well-being.

Before you begin the following Case Study

- Form cooperative groups with three to four students per group
- Select a team leader
- Assemble necessary materials
- Construct the sewer line
- Present the work

EXPLORATORY

Arrange an interview with a Civil Engineer or City Manager to discuss some of the waste treatment problems in your region. If your teacher arranges a classroom visit by such an engineer or manager, attend their presentation and participate in the discussion that follows. Ask questions about the history of your local sewage system and plans for future advancements.

INTERMEDIATE

Ask your teacher for the sheet that provides the "Sewer Layout Diagram." Pretend you are working for a construction company and you are looking at the engineering drawings for a proposed sewage line. Your job is to develop a layout for the location of each manhole so that construction of the sewer lines can begin. You must determine the angle at which each pipe enters the manhole, and the height of each manhole. This will require accurate measurements, assessment of what is needed for each manhole, and the preparation of a Take Off sheet for each of the manholes. If you follow the simple directions provided in the "To Do" section, you will successfully complete this project and be rewarded with a promotion.

ADVANCED

Once you have completed the Intermediate section of this project, you can begin constructing your sewer system. Using your calculations and the material provided (clay, straws, tape, and paper towel rolls), begin the construction of each manhole. Inlet and outlet angles are important in constructing the base. Follow your calculations, so the sewer system will allow sewage coming into the system to flow properly to the treatment plant.

After completing this part of your model, use brown cardboard or stiff paper to indicate what the contour of the ground might be from one manhole cover to the next. You should make the slope of the ground between the manhole covers as gradual as possible. What additional information do you need to make this model more realistic?

MATERIALS:

EXPLORATORY

None

INTERMEDIATE

None

ADVANCED

- Modeling Clay
- Straws
- Masking Tape
- Paper Towel Rolls
- Scissors
- X-acto Knife

Intermediate Challenge/Activities

1. For the problem, obtain the Sewer Layout Diagram from your teacher. No dimensions are provided. With your ruler, determine the distance between each manhole. Measurements should be rounded to the nearest quarter- inch (1/4", 1/2", 3/4", or one whole inch). Using the horizontal scale of 1" = 40 feet, convert each measurement to feet and mark the distance on the drawing. (For example, refer to the distance between manholes #1 and #2. This distance is 2" which is equal to 80 feet.)
2. Determine how far below the ground the sewer line is at each manhole. To start, you need to know the the sewer line depth at manhole #1. That information is provided as 30 feet (30'). You can now determine the depth of each manhole and the bottom measurement for each. To do this, you will need to allow for the downhill slope of the sewer line. (In this case, you will use a slope of 2 percent.) You can use the formula that indicates the slope will equal the amount of rise (or fall) divided by the horizontal distance. Written in symbols, the formula is S = R/HD, where S equals slope, R equals rise, and HD equals the horizontal distance. In this case, we will use a slope of .02 or 2 percent. (For example, refer to the distance between manhole #1

and manhole #2, which is 80 feet. A 2 percent slope over 80' equals 16 feet in fall.) Substitute the numbers you have for the symbols in the formula (S = 2 percent or 0.02, and HD = 80 feet), and calculate the fall. Do you agree that the fall is 0.16 of a foot?

3. With your ruler, measure the height of each manhole to the nearest quarter-inch and mark this on the drawing. Convert this measurement to a decimal. Add this number to the bottom measurement. This new number represents your top measurement. Place the top measurement on top of each manhole.
4. Refer to the information sheets provided and obtain a copy of the Take Off sheet from your teacher. Use your protractor to measure where the inlet and outlet holes should be. Mark the degree measurement between all holes and indicate which holes are inlets and which are outlets. (Pay particular attention to the arrows on the lines of the planning drawings.)
5. Transfer the information that you know about each manhole (manhole #, top, bottom, and height) to the Take Off Sheet. This completes Step 1. Follow the next few steps on the Take Off Sheet, and the manholes are ready for the construction of the sewer line model.

ASSESSMENT

Your success on this Challenge will be based on your completion of the activities below. The evaluation of your performance will be your participation in the activity, the accuracy of your measurement and your model construction, and the performance of your design. Your teacher will help you understand how your performance will be graded.

EXPLORATORY

Conduct an interview with an engineer or city manager and/or participate in presentation. Develop questions for the speaker. Participate in the field trip if your teacher provides one.

INTERMEDIATE

Analyze the forms and drawings provided by your teacher and complete all worksheets. Determine all the distances and dimensions from the Sewer Layout Diagram and record them as directed. Calculate the dimensions and depth of each manhole based on the established slope of the sewer line. Determine the angles of the pipes entering and leaving the manholes and record them on a Take Off Sheet for each manhole.

ADVANCED

Follow the design and build a model sewer line. It should be sloped in accordance with the sewer diagram so sewage consistently flows in the proper direction. Under the supervision of your teacher, test your sewer line system by running a small amount of water through the lines. Give a presentation on your work.

RESOURCES

History of World Plumbing - Plumbing and Mechanical Magazine
http://www.theplumber.com/H_index.html

History of Plumbing (Roto-Rooter company)
http://www.rrsc.com/rr_history.html

Water Conditioning and Purification Magazine
http://www.wcp.net/current.htm

"Paris Sewers and Sewermen" by Donald Reid (Harvard University Press, 1991).

IDEAS is the result of a project funded by The Engineering Foundation and organized by two of the worlds major engineering societies, the American Society of Mechanical Engineers (ASME) and the American Society of Civil Engineers (ASCE). Great lakes press has been granted permission by ASME and ASCE for the use of IDEAS projects within this textbook

PRODUCTION

There are three major parts of production. They are scheduling, producing and controlling.

Routing

Production planners or industrial engineers have to determine the number of parts to be built and when those parts or products will be ready for consumers. These planners are responsible for routing the product through the facility. They determine the exact route that an item will travel though a plant when in production. This layout is done well in advance of any parts being made and can easily resemble a maze of lines in a diagram that explains this in detail. For example, if you were to make a pizza in your kitchen this is the process that would take place:

1. Locate resource for the pizza, usually from the grocery store.
2. Arrange financing for the product
3. Arrange for the transportation of the product to storage on site. This would involve you driving to and from the grocery store.
4. Store the product until it is ready to be used. Usually in your freezer.
5. Prepare machine for heating. (Turn on the stove)
6. Remove product from storage
7. Remove packaging from product
8. Wait for machine to heat to temperature
9. Place product in machine to be heated.
10. Wait 15 minutes
11. Locate tools to remove product from machine heating it. Usually hot pads
12. Remove product from machine
13. Stage product in a place that is safe and sanitary for cooling
14. Locate tools to cut product into pieces.
15. Locate sanitary items needed to serve product on
16. Bring serving items and staged cut product to same area
17. Separate product onto serving items
18. Distribute product

This is every step that is usually taken when making a frozen pizza. It is explained as if it were a procedure in a manufacturing or production plant, but it is an effective way to describe how complicated a simple process can be if it were to be done on a large scale.

Scheduling

Materials, workers facilities, and machinery are all items that have to be planned for in the scheduling process. In order to determine how many of a product will be made, production planners do several things. They forecast an estimated demand for the product. This may come in the form of a detailed market analysis or an evaluation of a competitive product. Customer orders also determine the minimum number of products to be made. Production planners might also have to allocate for parts or products that will be used by the business itself. For instance, if a company manufactured trash bags, they would obviously also need trash bags in the everyday course of their business. They would have to determine how many bags to add to manufacturing to supply their business with the bags.

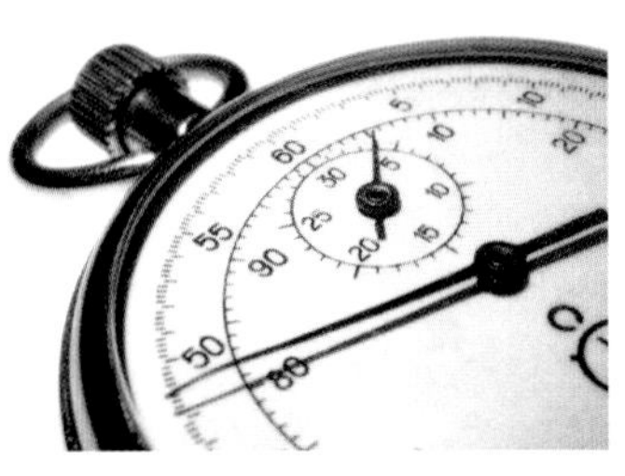

Lead time is another thing that a production planner has to consider in scheduling. Lead time is the amount of time it takes to get a particular part made, a new machine delivered to the plant, or have materials shipped to the production facility. In some startup businesses, lead time may also be an issue with the hiring of new personnel.

Just in time (JIT) manufacturing is a system that reduces the inventory of raw materials, parts or product components to the very minimum. These materials are scheduled to arrive at the production facility just as they are needed for use.

Controlling

Product level data reports are needed during production to assure a company that schedules from the plant, departments and workers are completing tasks on time. Changes in the way products are made usually come in small forms from an evaluation of the product level data. These reports not only catalog vital information for the current product, but are useful in the future development of other products.

Quality control is vital in the production of any consumer goods. Quality control involves inspecting materials, work products and labor, before and during production.

Materials have to be inspected when they arrive at the production facility. A quality product starts with quality materials. This doesn't necessarily mean high price. Quality materials can be defined as a material that meets the specification of a company to use for production. These materials may have to be tested for certain characteristics needed for the product. If the testing shows that the materials are not what are specified, the shipment of materials may be rejected. This may be as simple as a Rockwell C hardness test, or as complicated as a scanning electron microscope (SEM) test or tensile test.

Work in progress checks are made frequently during production. When a material is processed it has to be checked as it goes thought the production process. An example of this is when first stamping hoods for the MINI Cooper vehicle, plant personnel realized that the sheet metal that was being formed was tearing at the places where the hood was contoured. Engineer's initial discussion suggested that the counters had to be removed so that the tears could be eliminated. The designers of the vehicle and its contours in its hood felt this was unacceptable and that an alternative solution that maintained the hoods contours had to be found. After careful planning, the engineers provided specifications for the sheet metal to be thicker at the places that the contours were being stamped to insure that it would not tear. In the end the solution worked and satisfied not only the engineers, but the designers as the contours were maintained in the design of the hood. Quality control was an important part of this process. If the tears were not realized before the cars left the plants, it could have required an extensive, expensive recall of the vehicles to replace the hoods.

Work in progress checks produce three ratings on the status of a product. Rejects are parts that do not meet standards or cannot be repaired. Reworks are items that fail to meet standards, but can be repaired. Accepted parts are parts that meet standards that the company has set for that particular point in the process. Final products have to meet standards of the company that are initially set forth in the design process. If these standards have certain minimums (e.g. must support at least 10 pounds) then the final products have to meet the minimum standards to be allowed to go to the consumer. These minimum standards may be checked using machines, jigs and fixtures, or even the human eye for visual checks. If a product has to meet a minimum standard and is to be tested for that standard in its final form, it has to be destroyed. This process is called destructive testing. It may be that the part is put though multiple uses or split in half to check its internal features for flaws. These destructively tested parts are not sent to the customer, but discarded or documented as a part of a group of parts that have passed a test. Many times inspectors will label a part "inspected by" with a stamp or a sticker to assure the customer that it has been through the process.

Student Learning Activity

Systems poster

CASE STUDY

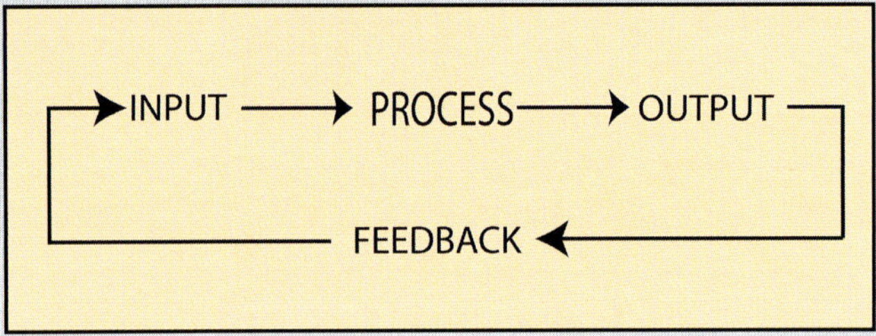

There are four parts of a system: Input, Process, Output, and Feedback. Your class will discuss a sample system poster and a grade may be assigned for student participation.

You are to work individually on your poster and choose a simple system for which you will provide pictures and text to explain each part of the system. Sketches, cut-out pictures, and Internet resources, may be used to find your information. Clarity is essential in this exercise. Your text should be neat and clean, and your pictures should illustrate what the text describes.

III OUTPUT

An output can be many different things. It might be products, structures, messages, people moved from one place to another, goods or services. These outputs are direct results or anticipated outcomes from what is to be designed. If a company wanted to provide homes for families in an undeveloped area of a state, they would go through many different processes in order to complete that goal. In the course of doing so they would produce a final product and many unwanted outputs such as scrap wood, metal, and many other wastes associated with their product.

Pollution is also another large unwanted output. Pollution can come in many forms. Chemical by-products are a large problem for companies that make film for cameras. These chemical companies have to spend millions of dollars each year to dispose of these chemicals properly.

Noise pollution is generated by construction companies that are building the new homes for people. While they may not have to pay because they pollute noise into the neighborhoods, they do have to follow city rules with regards to the hours that they operate their equipment and work on a site. Air and water pollution are significant problems for our country. Many companies that produce unwanted outputs have to also spend money on systems to clean the air and the water they will put back into the environment.

IV FEEDBACK

Feedback generates information or data that allows the user, consumer product, system or company to make informed decisions about what comes next. This could be used by a system to regulate itself at some set time interval.

For example if the temperature in your car reaches a certain temperature, a mechanical part (thermostat) is opened in order to allow fluid to flow to other parts of the engine. If this feedback information is ignored, or the mechanical part does not respond to the feedback, parts and systems in the car can fail. Similarly in that same vehicle, there are many feedback gauges that allow the driver to monitor what is going on in the car. Computers can adjust and monitor levels based on feedback from a system automatically. Nuclear power plants, space agency vehicles and monitoring systems and interactive homes are examples of where feedback is essential.

Optimization can be defined as doing the most with the least. A crude but distinct example would be that of cutting a 2" by 2" square out of a piece of paper. Would cutting it out of the center be the best solution? Or might it be a better solution to use the edges of the paper in a corner to eliminate two unnecessary cuts and have a large portion of the paper uninterrupted for future use? Optimization can save companies millions of dollars if done correctly.

Most design problems we face have no definitive answer. Most solutions are a series of compromises, which in turn allow the product to function well. An engineer or technician must assess all the parts of a problem and optimize each before the job is resolved. Webster's defines optimization as "The procedure or procedures used to make a system or design as effective or functional as possible, especially the mathematical techniques involved."

Chapter 5

Teamwork

Introduction

More than ever, schools are requiring their students to work in teams. These teams take various forms: joint study groups, laboratory groups, design groups in class; and design groups participating in extracurricular competitions.

This team emphasis in engineering study reflects the use of teams in the work world. Engineering employers are the principal drivers for the use of teams in schools. They seek engineering students who have team skills as well as technical skills. The purpose of this chapter is to communicate a team vision, to show why teams and cooperation are important to organizational and technical success, and to give practical advice for how to organize and function as a team. If you study this chapter, work through the exercises, and examine some of the references, you should have most of the tools you need to be a successful team member and team leader in school. Apply the principles in this chapter to your team experiences in school, and learn from and document each principle. By doing this, you will have a portfolio of team experiences that will be a valuable part of your résumé. Above all, you will have the satisfaction that comes from being a part of great teams. Being a part of a great team can be fun.

WHY DO CORPORATIONS FOCUS ON TEAMS?

A full-page advertisement in USA Today in 1991 noted the highly competitive climate in industry and what Chrysler (now DaimlerChrysler), for example, was doing about it. The use of interdisciplinary teams was an important part of the solution:

> "No more piece-by-piece, step-by-step production. Now it's teams. Teams of product and manufacturing engineers, designers, planners, financing and marketing people together from the start...It's how we built Dodge Viper from dream to

> showroom in three years, a record for U.S. car makers. From now on, all our cars and trucks will be higher quality, built at lower cost, and delivered to the market faster. That's what competition is all about."

The greater the problem, the greater the need for teams. Individuals acting alone can solve simple problems, but tough problems require teams. This section lists and then discusses the reasons many organizations have embraced the use of teams.

Why are teams so popular today?

- Engineers are asked to solve complex problems.
- More factors must be considered in design than ever before.
- Teams provide for greater understanding through the power of collaboration.
- Many corporations are international in scope, with design and manufacturing engineering operations spread across the globe.
- Since "Time to Market" is extremely important to competitive advantage, "concurrent engineering," inherently a team activity, is widely employed.
- Corporations increasingly are incorporating project management principles.

INCREASING COMPLEXITY OF PROJECTS

Engineers are asked to solve increasingly complex problems. The complexity of mechanical devices has grown rapidly over the past two hundred years. David Ullman, in his text "The Mechanical Design Process," gives the following examples. In the early 1800s, a musket had 51 parts. The Civil War era Springfield rifle had 140 parts. The bicycle, first developed in the late 1800s, has over two hundred parts. An automobile has tens of thousands of parts. The Boeing 747 aircraft, with over 5 million components, required over 10,000 person-years of design time; thousands of designers worked over a three-year period on the project. Most modern design problems involve many individual parts and many subsystems (e.g., mechanical, electrical, thermal controls), each requiring specialists acting in teams.

Engineering designers must consider more factors than ever before, including initial price, life cycle costs, performance, aesthetics, overall quality, ergonomics, reliability, maintainability, manufacturability, environmental factors, safety, liability, and acceptance in world markets.

Satisfying these criteria requires the collective teamwork of design and manufacturing engineering, marketing, procurement, business, and other personnel. The typical new engineering student, being interested primarily in technical things, tends to believe that engineers work in isolation while always seeking the best technical solution. This is wrong. Engineering involves solving difficult problems and finding technical solutions while considering numerous constraints. Engineers make things happen; they are doers. Engineers often straddle boundaries in organizations. They are technical experts who must understand economics and the dynamics of the business enterprise.

THE INTERNATIONAL FACTOR

Many corporations are international in scope, with design and manufacturing engineering operations spread across the globe. These operations require teams to work together who may never physically meet. Instead, they often share data via electronic means. For example, American design engineers might collaborate on a design with Japanese engineers for a product that will be manufactured in China. Manufacturing personnel from the Chinese plant may also contribute to the project. As another example, software engineers in the United States, the United Kingdom, and India may collaborate on a software project around the clock and around the world. At any time during a 24-hour period, programmers in some part of the world might be working on the product.

THE NEED FOR SPEED

Concurrent engineering is widely employed to achieve better designs and to bring products to market more quickly. Time to Market is the total time needed to plan, prototype, procure materials, create marketing strategies, devise tooling, produce, and release a new product to market. In the traditional company, each of these steps is done serially, one at a time. It is a relay race, with only one person running at a time, passing the baton to the next runner. The design engineer "tosses a design over the wall" to a manufacturing engineer who may not have seen it. That engineer then "tosses it over the wall" to procurement personnel, and so on.

Concurrent engineering, on the other hand, is a parallel operation. Like rugby, all of the players run together, side-by-side, tossing the ball back and forth as they move down the field. Marketing, manufacturing, and procurement personnel are involved from the beginning of the design phase. The use of teams and technologies, such as Computer-Aided Design (CAD)/Computer-Aided Manufacturing (CAM), rapid prototyping, shared data, and advanced communications, have changed the engineering process. By using teams and CAD/CAM technologies while faced with fierce international competition, U.S. auto makers have cut the time needed to bring a new car to market down from approximately 5 years to fewer than 3 years in many cases.

The timely delivery of products to the marketplace is critical for companies to be profitable. Jack Welch, General Electric's respected former CEO, has said, "we have seen what wins in our marketplaces around the globe: speed, speed, and more speed." Xilinx Corporation has stated that its research shows that a six-month delay in getting to market reduces a product's profitability by a third over its life cycle. A Business Week article stated: "Reduce product development time to one-third, and you will triple profits and triple growth."

Developing new products is a key to growth and profits. During a recent year, Rubbermaid Corporation introduced 400 new products, more than one per calendar day. At least one-third of Rubbermaid's $2 billion in annual sales come from products which have been in existence fewer than five years.

Engineering students must understand the importance of speed without compromising quality. Delays have many negative consequences, so instructors who supervise teams in the classroom should provide incentives to students for early completion of team projects or, at least, penalties for failure to meet deadlines. Student teams should recognize the importance of speed, push themselves and their teams toward early completion of projects, and document their successes on their résumés or portfolios.

IDEAS buoyancy

CASE STUDY

The need to travel across oceans, lakes, and rivers has challenged civilizations since the earliest of times. As a result, people have designed all kinds of watercraft—from rafts to aircraft carriers—to meet various needs. The most important aspects of designing any such craft are buoyancy, stability, and a controllable method of propulsion.

Other aspects of a craft's design depend upon its purpose. For racing boats, speed is the highest priority, so designs that create the fastest boats are considered best. For transport ships that carry valuable cargoes, load-carrying capacity may be more important than speed.

Designers must find the best combination of speed, capacity, and suitability for specific types of cargo. The Clipper ships that carried cargo from China to England were built for speed as well as capacity because the ship that arrived back first in England would get the best price for its cargo. Today, huge oil tankers are designed more for capacity than speed, but, given the risk of oil spills, safety has become an important goal in their design.

CHALLENGE:

EXPLORATORY

Research and model a type of watercraft. Your model should be buoyant (float in water) and stable (remain upright). Explain the significance of the design in regard to hull shape and stability in the water.

INTERMEDIATE

Construct a craft out of paper that will carry a mousetrap across a small pool of water using only wind power. The reason for the mousetrap is that it will just barely float by itself. To win the competition, the craft must carry the mousetrap. You have 10 minutes to design and build your craft. The mousetrap can be disassembled but all parts must be in the craft when the contest begins and ends. The craft will be placed one inch from one end of a pool at the start. One selected team member may blow upon the craft to propel it across the pool. The time will start when the craft is placed in the water and will end when it touches the other side of the pool. Once in the pool, no one may touch the craft.

ADVANCED

With a teammate, research, design, and construct a model of an electric boat that can travel quickly along a twelve-foot course. You will be provided with a DC motor and three 1.5 volt AA batteries along with various other materials to accomplish your task.

MATERIALS:

EXPLORATORY

- One 12" x 12" sheet of poster board
- 8 ounces of modeling clay for ballast
- One roll of clear tape

INTERMEDIATE

- One standard mousetrap
- Two sheets of 8 1/2" x 11" paper
- Scissors
- One roll of transparent tape

ADVANCED

- 1 plastic grocery bag
- 1 DC motor
- 1 pc. 6" x 16" Pink Foam
- 1 pc. 1/8" x 4" x 12" Plywood
- 1/4" x 3/411 x 30" pine
- 1 drinking straw
- 1 pc. 10" phone wire
- 1/8" welding or brazing rod 12" long
- 1/16" welding or brazing rod 12" long
- 24" of 1/4" wood dowel
- 1 ink pen
- Assorted rubber bands
- Empty aluminum soda can
- Three 1.5 volt AA batteries

The following example of an integrated Science, Math and Technology (SMT) activity is provided to show possible SMT connections to the challenges introduced earlier.

HULL DESIGN CONSIDERATIONS

Boat hulls that carry freight are designed to be hydrodynamically efficient so they will be economically efficient. Design tradeoffs should be considered for the following factors at a minimum:

- nature of the waterway
- type of freight to be transported (solid, liquid or granular)
- travel speed desired

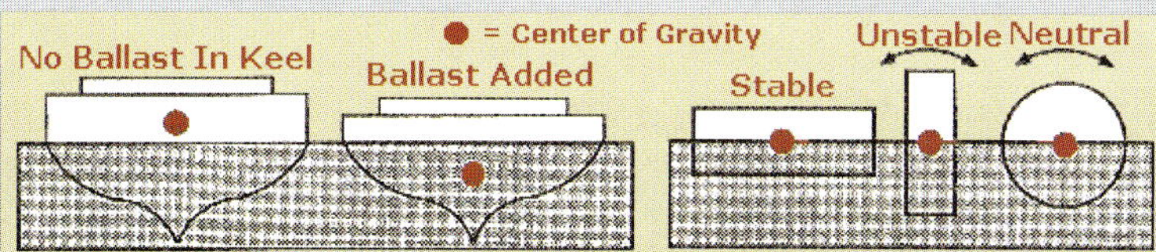

Fig. 5-A: Boat hull diagram

The ideal freight-carrying boat hull, in economic terms, would do the following:

- have the greatest possible payload
- be hydrodynamically designed
- permit the easiest and fastest motion while using the least amount of energy

HIGH-PERFORMANCE HULL CHARACTERISTICS

Three hull performance factors must be considered when designing a boat hull for a specific task: buoyancy, stability, and center of gravity.

Buoyancy has its effect when an object placed in a liquid pushes down on that liquid by force of gravity while the liquid is pushing up. The push upward is buoyancy. The object placed in the water will displace a volume of water equal to the weight of the object.

Stability is a characteristic of a vessel which can stay upright under all reasonable conditions. There are three states of stability: stable, unstable, and neutral. The force of gravity gives unstable objects a tendency to move forward to a stable position.

Center of Gravity is the focus point of gravitational pull in an object. The hull's center of gravity should always be low to improve stability. Ballast (additional weight) is usually added to a hull to lower the center of gravity.

Conduct an experiment on a boat model that has a hull design similar to the one shown in Figure 5-A. The model should be at least 18" long, 6" wide, and 4" deep. Build the model out of light materials,such as wood or plastic for the structure and sheets of plastic for the outside of the hull. Determine how much weight is needed as ballast to make the model stable so it floats without tipping over. Determine how far into the water the boat sinks before and after adding the ballast. Does it make any difference where you add the ballast in your model? Does the amount of ballast needed depend on whether you are carrying a heavy or light cargo or none at all? Compare your results with those of your classmates. Does a difference in the size of the boat model make a difference in the amount of ballast required to reach a state of stability?

PROBLEM-SOLVING PROCESS:

These steps may be helpful in approaching your activity:

- Form cooperative groups (two to three people).
- Research boat designs (books, Internet).

- Brainstorm for ideas.
- Sketch possible solutions.
- Decide how to construct, maneuver, and operate the model.
- Decide on and gather materials.
- Construct your design.
- Test and record data on your design.
- Present your work.

ASSESSMENT

Your success on this challenge will be based on your completion of the activities below. Three general criteria for evaluating your performance will be your participation in the activity, the accuracy of your measurement and your model construction, and the performance of your design. Your teacher will help you understand how your performance will be graded.

EXPLORATORY

Show a working knowledge of at least one boat design and incorporate this knowledge into your design. Use only the materials provided. Teacher approval is required before you may use any other material. The craft must be buoyant (float) and remain upright in the water.

INTERMEDIATE

Design and construct your craft within an allotted time-frame. Race your craft in the pool provided, following all rules.

ADVANCED

- Design, sketch, and build your craft.
- Test your boat to see if it will cross the pool powered solely by the motor and no more than two 1.5 volt batteries.
- Complete the construction of your model.
- Prepare and present your report.

RESOURCES

TIES Magazine, http://www.tiesmagazine.org/

Transportation, Energy & Power, Schwaller, A. Glencoe-McGraw Hill Publishing

Various Texts on Ships and Boats

World Submarine Invitational

PROJECT MANAGEMENT INCORPORATES TEAMWORK

Project Management is widely practiced in industries and government labs. Many engineers discover on their first job that they will frequently work in project-oriented teams. Unfortunately, most engineering students never take a project management class.

Project management principles were developed in the defense industry in the 1950s and 1960s as a way to manage Department of Defense contracts. The typical corporation is organized by vertical divisions or lines, where individuals are clustered by job functions. For example, a corporation may have separate research, manufacturing, engineering, human resources, product development, marketing, and procurement divisions. Project Management organizes individuals not by function but by products or projects. Selected individuals from research, engineering, manufacturing, human resources, procurement, and other divisions are gathered together as a team for a particular purpose, for example, solving a corporate problem, developing a new product, or meeting a crucial deadline. A Project Management approach, therefore, is inherently a cross-functional team approach and extremely useful for solving problems. Successful Project Management, however, can be quite challenging.

The Project Manager's job is to complete a project on time, within budget, and with the personnel they are given. Unfortunately, they are rarely able to hand pick their personnel, and their supervisors and/or customers typically set the project completion time and budget. In other words, project managers are rarely given all the people, time, and money they might prefer. These constraints are often mirrored in student design teams. Their teacher picks the team members, they do not have access to much money or resources, and their teacher determines the time frame.

ID	Task Name	Names	Start	Finish	Sep 2005 / Oct 2005 / Nov 2005 / Dec 2005 9/4, 9/11, 9/18, 9/25, 10/2, 10/9, 10/16, 10/23, 10/30, 11/6, 11/13, 11/20, 11/27, 12/4, 12/11
1	Preliminary Research	Everyone	9/14/2005	9/21/2005	
2	Site Visits	David Grundman, Ben Cox John Kronenwetter	9/14/2005	9/28/2005	
3	Problem Statement	Everyone	9/14/2005	9/21/2005	
4	Individual Brainstorming	Everyone	9/14/2005	9/28/2005	
5	Large Group Brainstorming	Everyone	9/21/2005	10/5/2005	
6	Analyze Potential Solutions	Everyone	9/28/2005	10/5/2005	
7	Preliminary Presentation	Everyone	9/20/2005	10/12/2005	
8	Poster Project	David Grundman, Brian Gerzsenyi, AJ Dahlberg	10/5/2005	10/17/2005	
9	Order Materials	AJ Dahlberg, Pat Flanagan	10/5/2005	10/26/2005	
10	Develop Prototypes	David Grundman, Brian Gerzenyi, AJ Dahlberg, Tyler Knoeck	10/12/2005	11/2/2005	
11	Test/Evaluate Designs	Shea Chyou, Patrick Flanagan	11/2/2005	11/9/2005	
12	Refine Design	Everyone	11/9/2005	11/23/2005	
13	Develop Final Product	Same as protoypes	11/16/2005	11/30/2005	
14	Final Presentation	Ben Cox, Houston Hoff, John Kronenwetter	11/21/2005	12/12/2005	

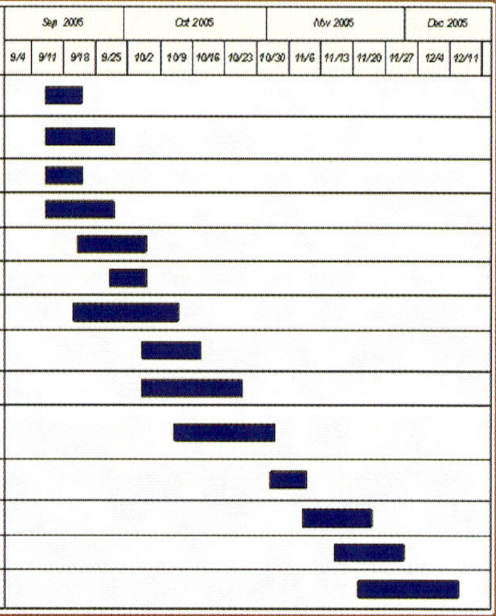

These conditions can be uncomfortable for you as students, but they prepare you for the actual engineering world.

Project managers use tools that any team leader would benefit from learning how to use. Project managers plan work requirements and schedules and direct the use of resources (people, money, materials, and equipment). One popular project management tool that student teams should learn to use is the Gantt Chart (shown at left). A Gantt Chart organizes and presents information on task division and sequencing. Tasks are divided into sub-tasks. The person(s) responsible for each task is listed.

WHAT MAKES A SUCCESSFUL TEAM?

Have you ever been a member of an excellent team? It is a great experience, isn't it? On the other hand, have you ever been part of an unsuccessful team? It was a disappointing experience, wasn't it? How do you measure or evaluate a team's success or failure? What specific factors make a team successful? What factors cause a team to fail?

First, it is important to define what a team is and is not. A team is not the same as a group. The term "group" implies little more than several individuals in some proximity to one another. The term "team," on the other hand, implies two or more persons who work together to achieve a common purpose. The two main elements of this definition are purpose and working together.

All teams have a purpose and a personality. A team's purpose is its task, the reason the team was formed. Its personality is the collective style, its people, and how the team members work together. Each team has its own style, approach, dynamic, and ways of communicating which are different from those of other teams. For example, some teams are serious, formal, and businesslike, whereas others are more informal, casual, and fun-loving. Some teams are composed of friends, and others are not. Friendship, though desirable, is an unnecessary requirement for a team to be successful. Commitment to a common purpose and to working together is essential. Regardless of style or personality, a team must exhibit professionalism. A team's professionalism ensures that its personality promotes productive progress toward its purpose.

TEAM ORGANIZATION

Organization is key for team success. Imagine if you were the boss in a cake-making factory. If your employees came to work in the morning and worked recklessly, would any quality cakes be made? Most likely no quality cakes would get made, if any were made at all. Organizing yourself and your team members to work together to accomplish a goal will make the process more efficient and less stressful. As was stated earlier in this chapter, teams are different from groups because teams are focused on a goal, and everyone participates to achieve that goal.

Your group of four students will be assigned an area of the room to organize a deck of cards, after they are dropped on the floor, into suits and sequential order. You will be allowed two attempts, and you should time how long it takes for each attempt.

TEAM ATTRIBUTES

The successful team should have the following attributes:

1. Common goal or purpose: ...team members are individually and collectively committed to that purpose.
2. Leadership: ...though one member may be appointed or selected as the team leader, ideally every team member should contribute to the leadership of the team.
3. Contributions: Each member makes unique contributions to the project. A climate exists in the team that recognizes and appropriately utilizes the individual talents/abilities of the team members.
4. Effective team communication: Regular, effective meetings; honest, open discussion; ability to make decisions.
5. Creative spark: Excitement and creative energy. Team members inspire, energize, and bring out the best in one another. A "can do" attitude. This creative spark fuels collaborative efforts and enables a team to rise above the sum of its individual members.
6. Harmonious relationships among team members: Team members are respectful, encouraging, and positive about one another and the work. If conflicts arise, peacemakers are on the team. The team's work is productive and fun.
7. Effective planning and use of resources: This involves an appropriate breakdown of tasks and effective utilization of resources (e.g., people, time, money).

TEAM MEMBER ATTRIBUTES

Individual team members of a successful team should have the following attributes:

1. Attendance: Attends all team meetings, arriving on time or early. Dependable, like clockwork. Faithful, reliable. Communicates in advance if unable to attend a meeting.
2. Responsibility: Accepts responsibility for tasks and completes them on time, needing no reminders or motivating. Has a spirit of excellence, yet is not overly perfectionistic.
3. Ability: Possesses abilities the team needs, and contributes these abilities fully to the team's purpose. Does not withhold self or draw back. Actively communicates at team meetings.
4. Creativity, Energy: Acts as an energy source, not a drain. Conveys a sense of excitement about being part of the team. Has a "can do"

attitude about the team's task. Has creative energy and helps spark the creative efforts of everyone else.

5. Personality: Contributes positively to the team environment and personality. Has positive attitudes, encourages others. Acts as a peacemaker if conflict arises. Helps the team reach consensus and make good decisions. Helps create a team environment that is both productive and fun. Brings out the best in the other team members.

HOW EFFECTIVE TEAMS WORK

Every team makes decisions to accomplish its work. The ability to make high-quality decisions in a timely manner is a mark of a great team. The process of how decisions are made greatly affects the quality of these decisions. Unfortunately, most teams arrive at decisions without knowing how they did it. In this section, we present ways in which decisions typically are made and discuss their advantages and disadvantages. Though it is recommended that a team decides in advance what decision-making process it will use, the best teams use a variety of means of making decisions depending on the circumstances. The following classification of ways teams make decisions is summarized from a section in the book Why Teams Don't Work, by Harvey Robbins and Michael Finley.

MODES OF TEAM ACTION

Consensus: A decision by consensus is a decision in which all the team members find common ground. It does not necessarily mean a unanimous vote, but it does mean that all have an opportunity to express their views and to hear the views of others. The process of open sharing of ideas often leads to better, more creative solutions. Unfortunately, achieving consensus can take time and may become unwieldy for larger teams.

Majority: Another way to make a decision is by majority vote. Whatever position receives the most votes wins. The advantage of this approach is that it takes less time than reaching consensus. Its disadvantage is that it provides for less creative dialogue than consensus, and there always will be a faction, the minority, who lose the vote and may become alienated.

Minority: Sometimes a small subset of the team, an elected subcommittee for example, makes the decision. The advantage of this is that it may expedite the decision. The disadvantage is that there is less overall team communication, and some team members may be prevented from making contributions to the decision. And with less input, an inferior decision may be reached.

Averaging: Averaging is compromise in its worst form and is the way Congress and some committees arrive at decisions. Averaging is often accomplished with haggling, bargaining, cajoling, and manipulation. Usually, no one is happy except the moderates. The advantage of averaging is that the extreme opinions tend to cancel one another out. The disadvantages are that there often is little productive discussion (unlike a consensus), and that the least informed can cancel the votes of the more knowledgeable.

Expert: When facing a difficult decision, there is no substitute for expertise. If there is an expert on the team, he or she may be asked to make a decision. If a team lacks an expert on a particular issue, which is often the case, the best teams recognize this and seek the advice of an outside expert. The advantage of this is that, theoretically, the decision is made with accurate, expert knowledge. The disadvantage of this is that it is possible to locate two experts who, when given the same information, disagree on the best course of action. How is a team assured that their expert gives them the best advice? Team members may be divided on which expert to consult, as well as on their assessment of the expert's credentials.

Authority Rule without Discussion: This occurs when a strong leader makes decisions without discussing the details with, or seeking advice from, the team. This works well with small decisions, particularly of an administrative nature; with decisions that must be made quickly; and with decisions to which the team is not well qualified to contribute , anyway. There are many disadvantages to this approach, however. The greatest is that the team's trust in its leader will be undermined. They will perceive that the leader does not trust them and wants to circumvent them. If one person is continually taking it upon himself to make all the decisions, or if the team is abrogating its responsibility to act together and, by default, forcing the decision onto an individual, then this is merely a group and not a true team. It is a loose aggregation of individuals.... a dysfunctional team.

Authority Rule with Discussion: This can be an effective way of making decisions. It is understood from the outset that the final decision will be made by one person, the leader or a delegated decision maker, but the leader first seeks team input. The team meets and discusses the issue. Many viewpoints are heard. Now, more fully informed, the appointed individual makes the decision. The advantage of this method is that the team members, being made part of the process, feel valued. They are more likely to be committed to the result. This type of decision-making process requires a leader and a team with excellent communication skills, and it requires a capable leader who is willing to make decisions.

Student Learning Activity

Desert Survival

CASE STUDY

A planeload of Uruguayan rugby players crashed into the Andes Mountains in 1972. Some survived, and as the days passed, they realized the tragedy wasn't over: Their lives depended on their ability to adapt and survive with what little they had. There have been books and articles published, and movies made about this true event. The books and the movies tell the story, but when it is dissected into its simplest form, it is about teamwork and problem solving. They had more troubles to unravel than the average person living in civilization because they had to figure out how to survive. For 72 days, those who survived the crash had to plan their day with the next day in mind. They had to work together and work through their differences.

Many engineering teams will have differences, and it is in human nature to fight for your own opinions and "stick to your guns".

However, for a project to survive, or a group of people to survive, certain compromises must be made. It is how a team, engineering or otherwise, goes through this process that is the valuable lesson to be learned.

> "The key elements in the art of working together are how to deal with change, how to deal with conflict, and how to reach our potential...the needs of the team are best met when we meet the needs of individual persons."
>
> Max DePree

To complete this case study, you will first work individually, and then come together in a team designated by your instructor. The situation is as follows:

It is approximately 10 AM in mid-August and you have just crash landed in the Sonora Desert in the southwestern United States. The light twin-engine plane containing the bodies of the pilot and co-pilot has completely burned. Only the airframe remains. Neither you nor the other survivors has been injured.

The pilot was unable to notify anyone of your position before the crash. However, he had indicated before impact that you were 70 miles from a mining camp, which is the nearest known habitation, and that you were approximately 65 miles off the course that was filed in your VFR Flight Plan.

The immediate area is quite flat and, except for occasional barrel and saguaro cacti, appears to be rather barren. The last weather report indicated that the temperature would reach 100 degrees that day, which means that the temperature at ground level will be 130 degrees. You are dressed in lightweight clothing: short-sleeved shirts, pants, socks, and street shoes. Everyone has a handkerchief. Collectively, your pockets contain $2.83 in change, $85.00 in bills, a pack of cigarettes, and a ballpoint pen.

Before the plane caught fire, your group was able to salvage the fifteen items listed in the next section. Your team's task is to rank these items in the order of their importance to your survival, starting with a "1" as the most important, to "15" as the least important, using the grid found in your workbooks.

Your team may assume the following:

1. The number of survivors is the same as the number in your assigned team.
2. You and your teammates are the people mentioned in the case study .
3. The team must stay together.
4. All items are in good condition.

The 15 Survival items:

- Flashlight
- Jackknife
- Sectional Air Map of the Area
- Plastic Raincoat (large)
- compress Kit with Gauze
- Magnetic Compass
- .45-Caliber Pistol (loaded)
- Parachute (red & white)
- Bottle of 1,000 Salt Tablets
- One Quart of Water Per Person
- Book Entitled "Edible Animals of the Desert"
- One Pair of Sunglasses Per Person
- Two Quarts of Vodka
- One Top Coat Per Person
- One Cosmetic Mirror

PART A

Each team member is to rank each item. DO NOT discuss the situation or the survival items until each member has finished the ranking the items on their own. You will have 15 minutes to complete the reading and the rankings.

PART B

After everyone has finished the individual ranking, rank the 15 items in order as a team. Once discussion begins, do not change your individual ranking. Your team will have 15 minutes to work on the collective rankings.

Add each of the members' individual total scores in Step 4 (the fourth column) and divide the total by the number of people in your group. This is your team's average individual score. Compare that number with the total number in Step 5. Was your group's average individual score close to the team score?

ASSESSMENT

Students will be asked to answer a series of questions based on their experience in the case study. The Total Quality Curriculum (TQC), from which this was adapted, has a series of class discussions and additional teaching materials to further this experience. The teacher should obtain these materials as well. Below are the questions students should answer. This one-day or two-day activity would warrant points given for the individual rankings, the team rankings, and the questions. Point values for each section should be relative to the amount of work for each section. For example: 10 points for the individual ranking, 15 points for the group ranking, and 20 points for the questions. Here are the student questions:

1. Did a leader in the group emerge?
2. How were decisions reached?
   a. Dictatorship
   b. Majority Rule
   c. Consensus
   d. Something else
3. Describe the styles of leadership the leader used.
4. Did disagreements occur?
5. How were disagreements handled?
6. Was everyone allowed to contribute?
7. Was there a conflict?
8. How was the conflict resolved?
9. Did everyone seem to agree about how the conflict was resolved?
10. How did the team members feel about the decision?

RESOURCES

http://www.skillsusa.org/tqcpage.html The Total Quality Curriculum (TQC) was designed to help meet the needs of American business and industry. These needs were brought about because of changes in the global economy. The American work force can no longer remain competitive using old methods. The TQC enhances SkillsUSA's Quality at Work movement by preparing students for the world of work in the classroom.

The Desert Survival Situation activity was adapted for use in the TQC by permission from J.C. Lafferty, Ph.D., Human Synergenics, Inc., 1987, San Diego, CA.

GETTING GOING IN A TEAM SETTING

If you have an opportunity to become part of a team, how should you approach it? What should your attitudes be?

1 Determine to give your best to help the team grow and to accomplish its purpose. Do not just tolerate the team experience.
2. Do not expect to have perfect teammates. You are not perfect and neither are they. Do not fret if your friends are not on your team. Make friends with those on your team.
3. Be careful about first impressions of your team. Some team members may seem to know all the right things to say, but you may never see them again. Others might not look like they have much to offer, but turn out to be the ones who attend all of the meetings, make steady contributions, and ensure the team's success. Be careful about selecting a leader at the first team meeting. Sometimes, the person who seems at first to be the most qualified to lead will never make the commitment necessary to be the leader. Look for commitment in a leader.
4. Be a leader. If you are not the appointed leader, give your support to the leader and lead from your position on the team. Do not be scared off by team experiences in the past where the leader did all the work. Be one who encourages and grows other leaders.
5. Help the team achieve its own unique identity and personality. A team is a kind of a corporation, a living entity with special chemistry and personality. No team is just like any other team. Every team is formed at a unique time and setting with different people and a different purpose. Being part of a productive team that sparkles with energy, personality, and enthusiasm is one of the rewards of teamwork. Being part of a great team will be a great memory.
6. Be patient. Foster team growth, and give it time to grow. Teammates begin as strangers, unsure of one another and of the team's purpose. Watch for and help the team grow through the stages of Forming, Storming, Norming, and Performing.
7. Evaluate and grade yourself and your team's performance. Document its successes and failures. Dedicate yourself to be one who understands teams and who acts as a catalyst, one who helps the team perform at a high level. Note team experiences in your portfolio, especially times when you took a leadership role. Prepare to communicate these experiences to prospective employers.

THE CHARACTER OF A LEADER

Every successful human endeavor involving collective action requires leadership. Great teams need great leadership. Without leadership, humans tend to drift apart, act alone, and lose purpose. They may work on the same project, but their efforts, without synchronization and coordination interfere with, rather than build on, one another. Without coordinated direction, people become discouraged, frustrations build, conflicts ensue, money is wasted, and time schedules deteriorate. But the greatest loss is a loss of human potential. A failed team effort leaves a bitter taste not easily forgotten. There is the alternative, and it is the thrill of team success.

Leader Attributes

Leaders ensure that the team remains focused on its purpose and that it develops and maintains a positive team personality. They challenge and lead the team to high performance and professionalism. With one hand they build the team, and with the other they build the project. In support of these objectives, leaders must do the following:

1. Focus on the purpose: Help the team remain focused on its purpose.
2. Be a team builder: Leaders may actively work on some project tasks themselves, but their most important task is the team, not the project. They build, equip, and coordinate the efforts of the team so it can accomplish its purpose and succeed as a team.
3. Plan well and effectively utilize resources (people, time, money): Leaders effectively assess and use team member abilities.
4. Run effective meetings: They ensure the team meets together regularly and that meetings are productive.
5. Communicate effectively: Leaders effectively communicate the team's vision and purpose. They effectively praise good work and give effective guidance for improvement of substandard performance.
6. Promote team harmony by fostering a positive environment: If team members focus on one another's strengths instead of weaknesses, conflicts are less likely to arise. Conflict is not necessarily bad, however. The effective leader must not be afraid of conflict. He or she should view conflict as an opportunity to improve team performance and personality and to refocus it on its purpose.
7. Foster high levels of performance, creativity, and professionalism: Maintain a grand vision. Challenge team members to do the impossible, to think creatively, and to stimulate one another to high performance.

Every team needs all the leadership it can get. Most teams have a single individual who is voted or appointed as its leader. On the best teams, however, many members contribute leadership in ways that support and complement the appointed leader. Like the appointed leader, every member should focus on the team's purpose, build the team, plan, recognize the gifts of others, contribute to effective meetings, communicate well, promote a harmonious team environment, and be creative. Ideally, every member, together with the leader, should build the team with one hand and labor on the project with the other hand.

John Maxwell defines leadership this way: "Leadership is influence." All team members, from their positions on the team, have the opportunity to influence the team's performance. A team can have an outstanding leader, yet fail if the team chooses to use its influence to avoid following the leader, to undermine the leader, or to promote team disunity. Conversely, a team with a relatively weak appointed leader can be successful if the team members pitch in and use their influence to build the team. Ultimately, the entire team - the members and the appointed leader - must work together to build a great team. It takes teamwork to make a great team.

Chapter 6

Problem Solving

Introduction

Most people, when they are children, learn a few reliable ways to solve problems. Most fifth graders in math class know common problem-solving methods like "draw a picture," "work backward" or the "guess and check" (check in the back of the textbook for the correct answer, that is). Unfortunately, many people choose to stop adding new problem-solving tools to their mental toolbox. "If I don't know how to solve it, it's probably not worth solving anyway," they may think. Or, "I'll pass this problem over for now and get back to my usual tasks." These thoughts limit them. Engineers, by definition, need to be good at solving problems and making things. Therefore, filling your toolbox with the right tools and knowing how to use them are critical components to becoming a successful engineer.

The road to the solution can be as important as, if not more important than, finding the actual solution. As a student, you must discover the learning process for navigating that road. Though you may not become an engineer, everyone should learn to develop good problem-solving skills, especially given all the technological complexity of the 21st century.

PROBLEM SOLVING

The toolbox analogy is appropriate for engineering students. Practicing engineers are hired to solve many problems. Engineers fill different roles in their daily jobs. These roles can require different problem-solving abilities. Your analytic tools represent what is in your toolbox. Your creative skills represent how you handle your tools. Problem solving can be broken into Analytic and Creative aspects. Most students are more familiar with analytic problem solving where there is one correct answer. In creative problem solving, no single right answer exists.

To understand the difference between these two kinds of problem solving, examine the one function common to all engineering disciplines: design. In the Design chapter (Chapter 7), we will detail a design process with ten stages. Those ten stages are as follows:

1. Identify the problem/product innovation
2. Define the working criteria/goals
3. Research and gather data
4. Brainstorm/generate creative ideas
5. Analyze potential solutions
6. Develop and test models
7. Make the decision
8. Communicate and specify
9. Implement and commercialize
10. Prepare post-implementation review and assessment

By definition, design is open-ended and has many different solutions. The automobile is a good example. When you are riding in a car, count the number of different designs you see while driving on the road. All of these designs were the result of engineers solving a series of problems involved in producing an automobile. The design process as a whole is a creative problem-solving process.

The other method, analytic problem solving, is part of the design process. Step 5, analyze potential solutions, requires analytic problem solving. When analyzing a design concept, only one answer exists to the question "Will it fail?" Civil engineers designing a bridge might employ a host of design concepts, though at some point they must assess the loads the bridge can support. Either the design will work or it won't.

Most engineering classes in colleges provide many opportunities to develop analytic problem-solving abilities, especially in the first few years. Though most pre-college engineering classes provide fewer opportunities, work on this important skill. Improper engineering decisions can put the public at risk. For this reason, proper analytical skills must be developed.

As an engineering student, you will be learning math, science, computer, and hands-on skills that will allow you to tackle complex problems later in your career. These are critical and are part of the problem-solving skill set for analysis. However, other tools are equally necessary.

Most people rely on two or three methods to solve problems. If these methods do not yield a successful answer, they become stuck. Exceptional problem solvers learn to use multiple problem-solving techniques to find the optimum solution. The following is a list of possible tools or strategies that can help solve simple problems:

1. Look for a pattern.
2. Construct a table.
3. Consider possibilities systematically.
4. Act it out.
5. Make a model.
6. Make a figure, graph, or drawing.
7. Work backwards.
8. Select appropriate notation.
9. Restate the problem in your own words.
10. Identify necessary, desired, and given information.
11. Write an open-ended sentence.
12. Identify a sub-goal.
13. Solve a simpler problem.
14. Change your point of view.
15. Check for hidden assumptions.
16. Use a resource.
17. Generalize.
18. Check the solution and validate it.
19. Find another way to solve the problem.
20. Find another solution.
21. Study the solution process.
22. Discuss limitations.
23. Get a bigger hammer.
24. Sleep on it.
25. Brainstorm.
26. Involve others.

In analytic and creative problem solving, you have different methods for tackling problems. Have you ever experienced being stuck on a difficult math or science problem? Developing additional tools or methods will allow you to tackle more of these problems effectively, making you a better engineer.

ANALYTIC PROBLEM SOLVING

Given the importance of proper analysis in engineering and the design process, develop a disciplined way of approaching engineering problems. Solving analytic problems has been the subject of a great deal of research, which has resulted in several models. One of the most important analytic problem-solving methods that students are exposed to is the Scientific Method. The five steps in the Scientific Method are as follows:

1. Define the problem.
2. Gather the facts.
3. Develop a hypothesis.
4. Perform a test.
5. Evaluate the results.

In the Scientific Method, the steps can be repeated if the desired results are not achieved. The process ends when an acceptable understanding of the phenomenon under study is achieved.

In the analysis of engineering applications, a similar process can be developed to answer problems. The advantage of developing a set method for solving analytic problems is that it provides a discipline to help young engineers when they are presented with larger and more complex problems. Just as a musician practices basic scales to set the foundation for complex pieces to be played later, an engineer should develop a sound fundamental way to approach problems. Fortunately, early in your engineering studies, you will be taking many science and math courses that will be suited to this methodology.

The analytic method we will discuss has six steps:

1. Define the problem and make a problem statement.
2. Diagram and describe.
3. Apply theory and equations.
4. Simplify the assumptions.
5. Solve the necessary problems.
6. Verify accuracy to required level.

Following these steps will help you better understand the problem you are solving and allow you to identify areas where inaccuracies might occur.

Step 1: Problem Statement

Restate the problem you are solving in your own words. In textbook problems, this helps you understand what you need to solve. In real-life situations, this ensures you are solving the correct problem. Write down your summary and check that your impression of the problem matches the original problem. Putting the problem in your own words is an excellent way to focus on the part of the problem you need to solve. Often, engineering challenges are large and complex, and the critical task is to understand what part of the problem you need to solve.

Step 2: Description

The next step is to describe the problem and list all that is known. In addition to restating the problem, list the information given and what needs to be found. This is shown in Example 6.1, which follows this section. Typically, in textbook problems, all the information given is what is needed for the problem. In real problems, more information is typically available than is needed to do the calculations. In other cases, information may be missing. Formally writing out what you need and what is required helps you to clarify this.

Drawing a diagram or sketch of the problem helps you understand it. Pictures help many people to clarify the problem and what is needed. They are a great aid when explaining the problem to someone else. The old saying could be restated as "A picture is worth a thousand calculations."

Step 3: Theory

State explicitly the theory or equations needed to solve the problem. You must write this out completely at this step. You will find that most real problems and those you are asked to solve as a college engineering student will not require exact solutions to complete equations. Understanding the parts of the equations that ought to be neglected is vital to your success.

Step 4: Simplifying Assumptions

As mentioned above, engineering and scientific applications often cannot be solved precisely. Even if they are solvable, determining the solution might be too costly. For instance, an exact solution might require a high-speed computer to calculate for a year to get an answer, and this would not be an effective use of resources. Weather prediction at locations of interest is such an example.

To solve the problem presented in a timely and cost-effective manner, simplifying assumptions are required. Simplifying assumptions can make the problem easier to solve and still provide an accurate result. Write the assumptions and how they simplify the problem. This documents them and allows the final result to be interpreted in terms of these assumptions.

Though estimation and approximation are useful tools, engineers are concerned with the accuracy and reliability of their results. Approximations are often possible if assumptions are made to simplify the problem at hand. An important concept for engineers to understand in such situations is the Conservative Assumption.

In engineering problem solving, "conservative" has a non-political meaning. A conservative assumption is one that errs on the safe side. We mentioned that estimations can be used to determine the bounds of a solution. An engineer should be able to look at those bounds and determine which end of the spectrum yields the safer solution. By selecting the safer condition, you are assuring that your calculation will result in a responsible conclusion.

Approximation can improve decision making indirectly as well. By using it to resolve minor aspects of a problem, you can focus on the more pressing aspects.

Engineers are faced with these types of decisions every day in design and analysis. They must determine what problem is to be solved and how to solve it. Public safety is paramount in the engineering profession. You might be faced with a short-term safety emergency where you need to prioritize solutions. You might be required to design a consumer product where you have to take the long-term view.

Step 5: Problem Solution

Now the problem is set up for you to perform the calculations. This might be done by hand or by using a computer. Learn how to perform simple calculations by hand. However, computer applications simplify complex and repetitive calculations. When using computer simulations, develop a means to document what you have done in deriving the solution. This will allow you to find errors more quickly, as well as to show others what you have done.

Step 6: Accuracy Verification

Engineers work on solutions that can affect the livelihood and safety of people. The solution an engineer develops must be accurate.

Engineers are responsible for verifying that their own solutions are accurate. Therefore, the student must develop skills to meet any required standard.

There are many different ways to verify a result. Here are some ways:

1. Estimate the answer.
2. Simplify the problem and solve the simpler problem. Are the answers consistent?
3. Compare with similar solutions. In many cases, other problems were solved similarly to the one in question.
4. Compare to previous work.
5. Ask a more experienced engineer to review the result.
6. Compare to published literature on similar problems.
7. Ask yourself if it makes sense.
8. Compare to your own experience.
9. Repeat the calculation.
10. Run a computer simulation or model.
11. Redo the calculation backwards.

Determining if your answer is correct may be difficult, but you should be able to tell if it is close or within a factor of ten. If you can tell that your answer is way off, you know that your approach to solving the problem is defective. Being able to back up your confidence in your answer will help you make better decisions on how your results will be used.

Estimation

Estimation is a problem-solving tool that engineers need to develop. It can provide quick answers to problems and to verify complicated analyses. Young engineers are invariably amazed when they begin their careers at how the older, more experienced engineers can estimate the solution to a problem so effectively.

Though estimation or approximation may not yield the precision required for an engineering analysis, it is a useful tool. One thing that it can do is help check an analysis. Estimation can provide bounds for potential answers. This is especially critical with today's dependence on computer solutions. Your method may be correct, but one mistyped character will throw your results off. Have confidence in your results by developing tools to verify accuracy.

Case Study: Book Smart

The primary purpose of this case study is to develop measuring and averaging skills. You will be required to find the average paper thickness used in three school textbooks. This answer will be compared to the actual paper thickness measured with a micrometer.

Your teacher will divide you up into teams of two and should assign you a Vernier caliper and a micrometer. Use the Vernier caliper first to total thickness of each book's paper content (measure to the nearest .001 inch). To find the thickness of a page, you must divide the number of pages by two since there are two sides to every page. Record each of your answers in separate columns for each book. Average your three books together to find a final average for your group. Turn in your results to your teacher or share them with the class. Calculate the class average for each book's paper thickness. You should produce one final number for this step. Compare your group average to the class average. Which group was closest to the class average?

Now use a micrometer to measure the thickness of one page of each book. Multiply the thickness of the page by the number of pages in the book and divide by two. How does this number compare with the Vernier caliper average that you measured in the first part of this study? Average your group's micrometer-gauged pages together to get a final average thickness. Again, turn this in to the teacher or present the information to the class so you can make comparisons again.

Plot the resulting answers on a spreadsheet for each of the sections (Vernier and micrometer measurements) and compare them to the class's average measurements.

Creative Problem Solving

Many engineering problems are open-ended and complex. Such problems require creative problem solving. To maximize the creative problem-solving process, a systematic approach is recommended. As with analytic problem solving, developing a systematic approach to using creativity will pay dividends in better solutions.

Revisiting our toolbox analogy, the creative method provides the solid judgment needed to use your analytic tools. Individual thinking skills are your basics here. The method we will outline will help you to apply those skills and choose solution strategies effectively.

There are numerous ways to look at the creative problem-solving process. We will present a method which focuses on answering these five questions:

1. What is wrong?
2. What do we know?
3. What is the real problem?
4. What is the best solution?
5. How do we implement the solution?

By dividing the process into steps, you are more likely to follow a complete and careful problem-solving procedure, and more effective solutions will result. This allows you to break a large, complex problem into smaller simpler problems where your various skills can be used. Using such a strategy, the sky is the limit on the complexity of projects an engineer can complete. Seeing what can be done is astonishing and satisfying.

Sample Problem: Let's Raise Those Low Grades!

To illustrate the problem-solving process, consider an example of a problem and work through the process. The problem we will use for illustration is a student who is getting low grades. This certainly is a problem that needs a solution. Examine how the process works.

What's Wrong?

In the first step of the problem-solving process, an issue is identified. This can be something stated for you by a supervisor or an instructor, or something you determine on your own. This is the stage where entrepreneurs thrive, looking for an opportunity to meet a need. Similarly, engineers look to find solutions to meet a need. This may involve optimizing a process, improving customer satisfaction, or addressing reliability issues.

To illustrate the process, take our example of the initial difficulty that beginning engineering students can have with grades. Most students were proficient in previous grades and might not have had to study much. The demands of engineering programs can catch some students off guard. Improving grades is the problem we will tackle.

What Do We Know?

The second step in problem solving is the gathering of facts. All facts and information related to the problem identified in the first step are gathered. In this initial information gathering stage, do not evaluate if the data are central to the problem. As will be explained in the Brainstorming and Idea Generating sections in Chapter 7, premature evaluation can be a barrier to generating sufficient information.

With our example of the student questioning his grades, the information we generate might include the following:

1. Current test grades in each course
2. Current homework and quiz grades in each course
3. Percent of the semester's grade already determined in each course
4. Class average in each course
5. Instructor's grading policies
6. Homework assignments behind
7. Current study times
8. Current study places
9. Effective time spent on each course
10. Time spent doing homework in each course
11. Performance of friends
12. Previous school grades
13. Previous school study routine

Part of this process is to list the facts you know. To be thorough, you should request assistance from someone with a different point of view. This could be a friend, an advisor, a parent, or an instructor. They might come up with a critical factor you have overlooked.

What Is the Real Problem?

This stage is often skipped, but it is critical to effective solutions. The difference between this stage and the first is that this stage of the process answers the question "why?" Identifying the initial problem answers the question of "What is wrong?" To fix the problem, a problem solver needs to understand why the problem exists. The danger is that only symptoms of the problem will be addressed rather than root causes.

In our example, we are problem solving low class grades. That is the "What" question. To understand a problem, the "Why" must be answered. Why are the grades low? Answering this question will identify the cause of the problem. The cause is what must be dealt with for a problem to be addressed.

In our case, we must understand why the class grades are low. Assume that poor test scores result in the low grades. This does not tell us why the test scores are low. This is a great opportunity for brainstorming potential causes. In this phase of problem definition, do not worry about evaluating the potential causes. Wait until the list is generated.

Possible causes may include the following:

1. Poor test-taking skills
2. Incomplete or insufficient notes
3. Poor class attendance
4. Not understanding the required reading
5. Not spending enough time studying
6. Studying the wrong material
7. Attending the wrong class
8. Studying ineffectively (e.g., cramming the night before tests)
9. Failing to understand the material (might reveal need for tutor or study group)
10. Using solution manuals or friends as a crutch to do homework
11. Not at physical peak at test time (i.e., up all night before the test)
12. Did not work enough problems from the book

After you have created the list of potential causes, evaluate each as to its validity. In our example, there may be more than one cause contributing to the low grades. If so, make a rank-ordered list. Rank the causes in order of their impact on class performance.

Assume the original list is reduced to the following:

1. Poor class attendance
2. Studying ineffectively
3. Not spending enough time studying
4. Not understanding the material

Of these factors, rank them in order of their impact on grades. You may be able to do this yourself or you may need help. An effective problem solver will seek input from others when appropriate. In this case, an instructor or advisor might be able to help you rank the causes and help determine which ones would have the greatest impact. For this example, assume the ranking is as follows:

1. Not understanding the material
2. Studying ineffectively
3. Not spending enough time studying
4. Poor class attendance

We determined that while class attendance was imperfect, it was not the key factor in the poor performance. The key item was not understanding the material. This goes with ineffective studying and insufficient time.

What Is the Best Solution?

Once the problem has been defined, potential solutions need to be generated. This can be done by yourself or with the help of friends. In an engineering application, it is wise to confer with experienced experts about the problem's solution. This may be most productive after you have begun a list of causes. Experts can comment on your list and offer their own as well. This is a great way to get your ideas critiqued. After you gain more experience, you will find that technical experts help you narrow the choices rather than providing more. Also, go to more than one source. This may provide more ideas as well as help with the next step.

In our example, the technical expert may be an instructor or advisor who is knowledgeable about studying. Assume you have generated the following list of potential solutions:

1. Get a tutor.
2. Visit the instructor during office hours.
3. Make outside appointments with instructors.
4. Visit help rooms.
5. Form study group.
6. Outline books.
7. Get old exams.
8. Get old sets of notes.
9. Outline lecture notes.
10. Review notes with instructors.
11. Do extra problems.
12. Get additional references.
13. Make a time schedule.
14. Drop classes.
15. Retake the classes in the summer.
16. Go to review sessions.

The list must be evaluated, and the best solution decided upon. This convergent phase of the problem-solving process is best done with the input of others. It is especially helpful to get the opinions of those who have experience and/or expertise in the area you are investigating. In an engineering application, this may be a lead engineer. In our case, it could be an academic advisor, a teaching assistant, or an instructor. These experts can be consulted individually or asked to participate as a group.

Implementing the Solution

Implementing the solution may seem trivial and not worth discussing. This step, however, is a critical phase of the problem-solving process. In our scenario, the solution selected may have been to do extra problems.

To accomplish this solution, appropriate additional problems must be selected, done, and corrected. Implementation probably requires the assistance of the instructor to help select appropriate problems. To have the greatest chance of success, the student should get the assistance of the instructor. Assistance also could come from a study group.

In engineering applications, implementation can be a critical phase of the process. Most of the solutions to problems require additional resources such as money or the cooperation of other groups not directly affected by the problem or under your control. Early in your career, you will not have control over the people who need to implement solutions, so you will need to sell your ideas to the managers who do. An effective implementation plan will be critical to getting your solutions accomplished.

Evaluating the Solution

Implementation does not necessarily end the problem-solving process. Just as the design process is a circular process, with each design leading to possible new designs, problem solving also can be cyclic. Once a solution has been found and implemented, an evaluation should be performed. As with the other steps in the problem-solving process, this step begins when a problem solver asks what makes a successful solution and how it will be evaluated. To evaluate a solution, a measurement of (or a "criterion" for) success must be established. Sometimes, there are objective criteria for evaluation, but in many circumstances the criterion for success is more subjective.

Once the criterion has been established, the evaluation process should be defined, including who will evaluate the solution. Often it is desirable to get a neutral viewpoint from someone who was not involved in the formulation of the solution process as the evaluator.

If the solution is a success, the process may be complete. Often in engineering applications, solutions are intermediate and lead to other opportunities. The software industry is a prime example. Software is written to utilize the current computer technologies to address issues. Almost as soon as it is completed, new technologies are available which open up new opportunities or require new solutions.

Sometimes, a "success" does not address the true problem. In these cases, the evaluation process may require significant time before it can be determined a true success. In other cases, solutions have unintended outcomes that require a new solution even if the initial solution was deemed a success.

Whether a solution is effective or not, there is value in taking time to reflect on the solution and its implications. This allows you as a problem solver to learn from the process and the solution. Evaluating solutions and opportunities is a valuable skill for an engineer and is discussed further in the critical thinking section of this chapter.

Truss building

CASE STUDY

Trusses are used every day to build houses, bridges, towers and even above a stage to hang lights and wires. When learning about structures, two forces are easy to illustrate, tension and compression. Sheer and torsion are also important, but we must have a firm grasp of the terms compression and tension and how they interact together in structures.

Tension is the pulling apart of a beam or cable and compression is the force acting to shorten or compress a beam or cable. A bridge has to be able to resist both the forces of tension and compression (along with torsion) to be strong.

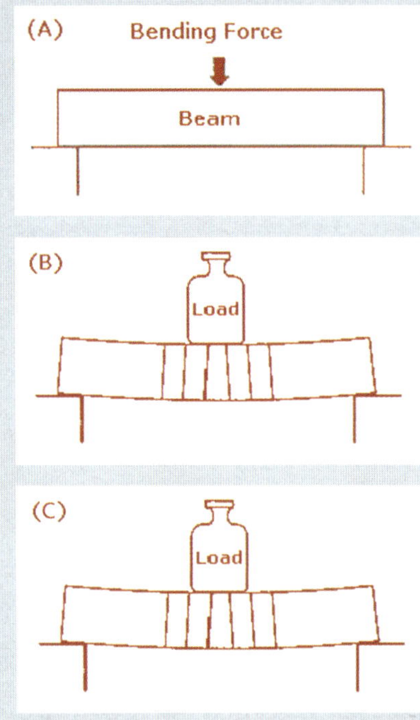

When a bridge is built using one or more beams, the bending forces that are applied by a load will be "across" the beam. In Figure A (left), the force is acting upon the member at right angles to its length. In such circumstances, the beam may be called a "member," and the action is called a bending force.

Use the foam materials provided by your teacher and draw the lines on the pieces as shown in Figure B (left). Place a small weight on the foam beam to create stress that will cause it to bend. Identify the part of the beam that is placed under tension (is being pulled apart), and which part of the beam is under compression (is being pushed together).

On beam (C), draw the lines along the length of the beam. These lines represent planes that run along and across the beam. Place the same weight used on beam (A) on beam (B). Can you determine which of the lines on the side of the beam gets shorter, which gets longer, and which remains about the same length? The line that remains the same length represents a neutral surface that remains unstressed. If a beam is loaded to the point that it will break, where will the break most likely be?

- At the end of these activities you will see a connection between simple geometry and the technology of building bridges.

- At the end of these activities you will be able to design simple wood structures that will not collapse when a load is applied to them.

- At the end of these activities you will have experience in adding steel cable and wood braces to strengthen a structure.

- At the end of this lab you will work as a team to figure out a way to combine parts from two groups into one bridge

- At the end of this lab you will work in a team to construct a 3-foot long bridge out of pieces of wood 18-inches long.

- At the end of these activities you will be able to design a long bridge from short material.

Please complete the Sections in this activity in your workbook:

Structure Lab #1, Exploring Tension and Compression
Structure Quiz
Structure Lab #2, From Triangles to Bridges
Structure Lab #3, Build a long bridge from short boards

MATERIALS

Four types of boards

- Regular 19" long bolt holes 18" center to center
- Regular green same as above but has extra holes drilled for bracing the truss bridge.
- Long 26", bolt holes 25" center to center
- Short 12", 11" center to center

Two types of cables

- Regular: will fit over two bolts 18" apart (a regular board)
- Long: will fit over two bolts 25" apart (a long board)

Materials to build one truss bridge:

This will supply a group of four with all the materials they need for all the labs. Bolts are all 1/4 inch.

Regular boards	18
Long boards	2
Cables long	4
Cables regular	4
Regular green	2
Short boards	4
Hex or wing nuts	18
Bolts 2"	4
Bolts 2 1/2"	8
Bolts 3"	4
Bolts 4 1/2"	2

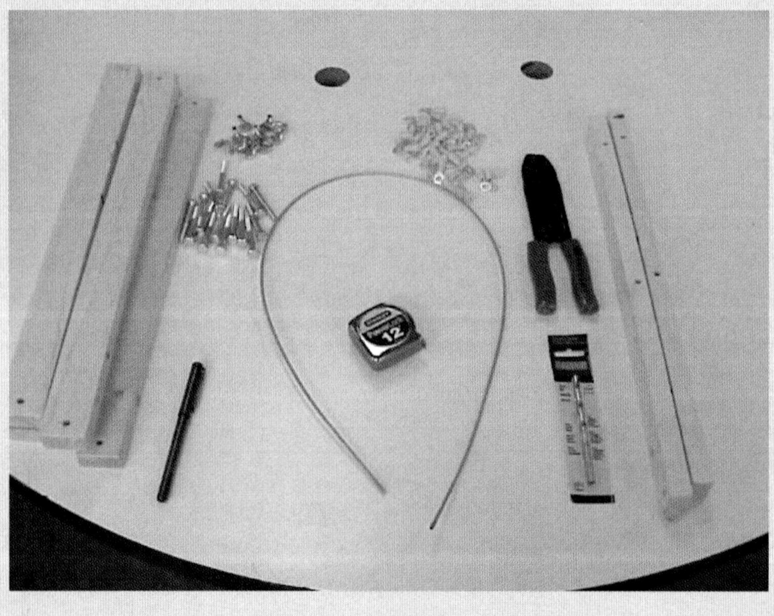

Chapter 7

Design and Modeling

What Is Engineering Design?

Orange County Choppers, the custom motorcycle shop popularized on the Discovery Channel show "American Chopper," uses SolidWorks® software to design the sweeping exhaust pipes, daring wheels, and other key features that make its street machines unique. SolidWorks' ease of use allows Orange County 's lead parts designer to convert cool ideas into 3D models without any prior computer-aided design (CAD) software training.

Engineers create things. Engineers build things. To perform these tasks, engineers must be involved in design or in a design process. So what is engineering design and an engineering design process? Webster's dictionary defines design as "to create, fashion, execute, or construct according to plan." It defines a natural process as "a natural phenomenon marked by gradual changes that lead toward a particular result... a continuous operation or treatment, especially in manufacture."

This chapter may seem closely related to Chapter 6 (Problem Solving). Indeed, there are many similar topics and themes in Chapter 6 that are important in design. One of the critical aspects which has an impact on design is the issue of external constraints that can influence the outcome of the process. (Appropriate references back to Chapter 6 will be noted throughout this chapter.)

Students who graduate with an engineering degree from an accredited program have had a significant amount of design experience as part of their education. The Accreditation Board for Engineering and Technology (ABET) has traditionally defined engineering design as follows:

> "Engineering design is the process of devising a system, component, or process to meet desired needs. It is a decision-making process...in which the basic sciences and mathematics and engineering sciences are applied to convert resources optimally to meet a stated objective. Among the fundamental elements of the design process are the establishment of objectives and criteria, synthesis, analysis, construction, testing, and evaluation. ....it is essential to include a variety of realistic constraints, such as economic factors, safety, reliability, aesthetics, ethics, and social impact."

Many beginning engineering students often confuse the design process with drafting or art-related work. To clarify this, ABET adds the following:

> "Course work devoted to developing computer drafting skills may not be used to satisfy the engineering design requirement."

THE DESIGN PROCESS

No single uniform approach to engineering design is followed by practicing engineers. Some firms approach engineering design as a short, simple process with only a few steps, and others use a more complex, multistep method with several stages. No matter what process is used, engineering design is always continuous. The completion of one design, or the solution of one problem, may serve to open opportunities for future designs or modifications.

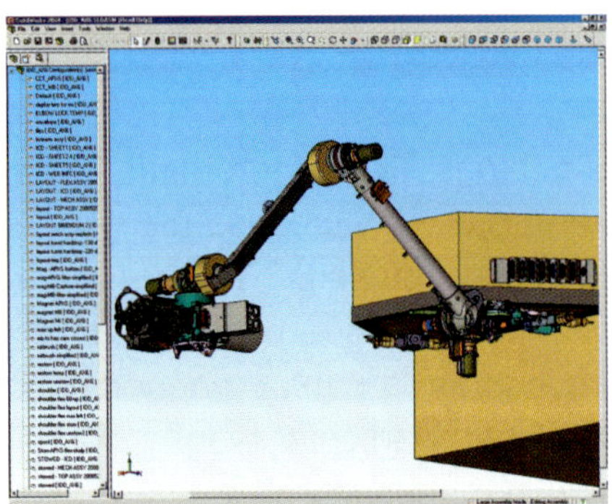

SolidWorks® Office provided an integrated set of engineering and design communication tools, which streamlined communications between ASI designers, scientists, and mission planners, enabling the design team to produce an innovative, robust design under a compressed design schedule.

A design process is used whether a product is being developed for an ongoing manufacturing process, where thousands or even millions of units of an item will be produced, or for a one-time design, such as with the construction of a bridge, dam, or highway exit ramp.

SIGNIFICANT FACTORS

Design is a process used by engineers to generate products, processes, and systems based on the recognition of a need. The following are significant factors in the design process that students should use in all case studies.

Functionality: The product or solution must fulfill its intended purpose. Imagine a table a student will use for studying. If a person were to stand on the table, would it support the weight? Was this table intended to support a person's weight? The table's original design was to support something other than a person's weight. The team that developed the table likely would not have intended it for the excessive weight.

Quality: The product or solution must be designed to meet certain minimum standards. The soles of shoes will be an example for quality. Students were asked how long they think their shoes will last. Most said that they use them for no more than two years. So why, if most of our day is spent walking or running on harder surfaces, such as concrete, are our shoes made of a softer surface like rubber? The answer is comfort.

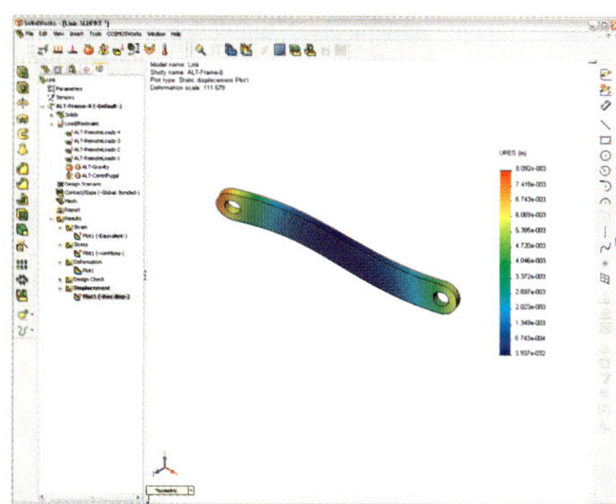

COSMOSMotion is fully integrated with embedded COSMOSWorks Designer analysis software. Based on how the mechanism moves COSMOSMotion will automatically transfer the loads and boundary conditions to COSMOSWorks Designer to perform a structural analysis.

When a worker once complained to his employer of being tired all the time and of having chronic back pain, he was sent to the doctor for evaluation. The doctor found the problem to be his shoes. Poor quality with improper support for his feet led to back pain and fatigue. Quality has a relationship to the conditions of an item's intended use. This spin-off of functionality is important because quality is defined according to the proper or improper use of the item. Are casual shoes intended to be used to play basketball in? Quality has to be evaluated within the context of intended use.

Safety: The product must be designed to comply with codes and regulations to provide safe user operation. Americans are becoming more safety conscious than ever, in everything from our cars to our homes. Air bags have become the norm in passenger vehicles since 1997; however, some have posed deadly problems. The force with which they are activated in an accident has injured many people, some of them fatally. These early air bags were sometimes activated not as the result of an accident. Occasionally, a driver's improper installation of a child car seat (facing the windshield) caused death to the child from the inflating force of the air bag. The first generation airbags have been redesigned many times. For example, a switch exists to turn off the passenger seat air bag. A driver now has the option to disable the air bag.

Ergonomics: The product must be designed so the user can operate it with ease and maximum efficiency. Ergonomics is sometimes called human factors engineering. Designing chairs for long-term comfort involves ergonomics. One famous fast food chain restaurant designs their chairs so that customers are comfortable only for the time it takes to eat. This way the customer will leave soon so employees can clean the table and chairs for the next people to keep business moving.

Ergonomic Task Chair, Carbon Design, Inc.

Imagine this scenario: a man and his family are going to be traveling east with three children, and they want to buy a conversion van. The man is 6'3" tall - much larger than the average person. His wife is 4'11" and petite. The man purchases the van and begins the trip east. He drives until fatigue sets in, and then turns the wheel over to his wife so he can get some rest. When he awakes from his nap, he notices that they are way behind schedule because she has been driving at 45 MPH on the expressway. When asked why she was going so slowly, she replies, "It was as fast as I could go since my legs are so short and my feet could not press any further down on the pedals."

The manufacturer had never considered that an adult only 4'11" in height would drive a full-sized conversion van. Manufacturers have begun to consider more in-depth ergonomic factors in their designs and their marketing plans.

Appearance: The appeal of a product to a customer may be based on the selection of materials, processes, finish, color, or shape. If you, as a potential consumer, do not like what a finished product looks like, you are less likely to purchase it. For example, a "new" baseball is colorful and clean, with no rips or tears. However, in some fashionable stores "new" baseball caps have torn, pre-formed, and dirty bills. People have different definitions of what is appealing or stylish. Companies invest millions of dollars in major marketing campaigns to get consumers to buy into their ideas. Ultimately, the consumer decides whether or not to purchase the product.

Environmental Considerations: The product must be designed so it does not adversely affect the environment. Many people immediately think of oil dripping from their cars and smog hovering over a city. Though these adverse conditions are included in this category, so are many others. Wind tunnels serve as an example. Planning commissions and city engineering departments continue to construct large buildings amongst large buildings and neglect to do any kind of wind analysis on the new structures or their surrounding environment. As a result, a five-mile an hour breeze could be increased by the configuration and heights of the buildings into a 15-mile an hour wind.

Economics: The product must be produced at the least cost without sacrificing safety or effectiveness. Engineers are challenged to keep costs down to increase company profit. On occasion, they may be tempted to take short cuts in production that lead to unsafe products. More often than not, the public is not concerned, nor do we know about these actions taken by companies, until something drastic happens.

For example, consider the Sam Poong department store in Japan that collapsed, killing 1,500 people. Store management left the building early in the morning after cracks appeared in the walls and ceiling, but they did not tell employees or customers of the danger. Shopping went on as usual throughout the building. Then the fifth floor collapsed onto the fourth, setting off a chain reaction all the way to the ground level. The building collapsed because of poor enforcement of building codes and too much water added to the concrete mix to lessen the costs.

STAGE DESIGN

This chapter will present one of the possible design processes -the ten-stage process. This process will be applied to an existing product to analyze how each of the ten stages may have been applied in the product's realization.

The ten stages that make up the process are as follows:

Stage 1:	Identify the problem/product innovation.
Stage 2:	Define the working criteria/goals.
Stage 3:	Research and gather data.
Stage 4:	Brainstorm/generate creative ideas.
Stage 5:	Analyze potential solutions.
Stage 6:	Develop and test models.
Stage 7:	Make the decision.
Stage 8:	Communicate and specify.
Stage 9:	Implement and commercialize.
Stage 10:	Perform post-implementation review and assessment.

The process begins with the appointment of a project manager or team leader. This individual will be responsible for oversight of the entire process to ensure that certain key elements of each stage have been satisfied before the project moves to the next stage. This person will be responsible for recruiting team members of varying backgrounds and expertise for each of the stages. The team will typically include non-engineers and engineers. Some of the team members will be used throughout the process, and others will be needed only for certain parts of the process.

Stage 1: Identify the Problem

Engineers are problem solvers, and the problems they solve often are identified as specific customer needs or problems. For example, a new artificial leg or arm may be required to overcome a particular handicap; increased gas mileage standards demand higher-efficiency engines; a new computer program is needed to monitor a modified manufacturing process; or new safety devices are required to better protect infants in automobiles. Therefore, the first stage of problem solving in engineering design is to establish and

Using SolidWorks® Office, ASI engineers developed the robotic arms for NASA's Mars Exploration Rover (MER) mission, which provided the first opportunity for detailed scientific studies on the Martian surface.

clarify the problem and to identify information sources to understand the problem's scope and nature.

The project manager will call upon the resources of various individuals to assist with these initial stages in the process. Many sources may be "outside" sources. Many firms have a research and development unit made up of scientists and engineers who have the training and expertise to assist with problem evaluation. In addition, sales engineers, who maintain consistent contact with outside individuals, can provide valuable input on problem identification. If the problem is one of expanding a current product line or modifying an existing system to improve it or make it better fit customer needs, management likely will play a role in the definition of the problem. Each group would be represented on the team at this stage.

Stage 2: Define the Working Criteria and Goals

Once the problem has been identified, the team must be able to validate it throughout the design process. This requires the establishment of working criteria, or standards, which can be used in each of the ten stages to measure possible solutions. The ultimate objective at this stage is to to establish preliminary goals which will act as the focal point for the team as it works through the process. The development of some working criteria provides a means to compare possible solutions. At this stage of the design process, everything is preliminary, so the team can still modify the criteria.

Examples of working criteria could include answers to the following questions:

1. How much will it cost?
2. Will it be difficult to produce?
3. What will be the size, weight, and strength?
4. What will it look like?
5. Will it be easy to use?
6. Will it be safe to use?
7. Do any legal concerns exist?
8. Will it be reliable and durable?
9. Can it be recycled?
10. Is this what the customer truly wanted?
11. Will our customers want to purchase it?
12. Will customers want to purchase this version instead of a competitor's product?

Once some preliminary working criteria have been established, overall goals for the process are developed. These are a statement of objectives which can be evaluated as the design process evolves. Using the example of new standards implemented to increase gas mileage and reduce emissions in automobiles, the goals for the design might be "To develop an automobile engine which produces 25 percent less emissions while increasing gas mileage by 10 percent."

Having overall goals established for the project provides a means of evaluating, monitoring, and changing, if necessary, the focus of the process as it progresses through the ten stages. For the project manager, the criteria and goals become a "checkpoint" for assessing the progress to date, and will help them determine if the project is ready to move to the next stage or if the process needs to return to Stage 1 for re-evaluation.

Stage 3: Research and Gather Data

This stage is important to all remaining stages of the design process. Having good, reliable background information is necessary for the team to begin exploring all relevant aspects of the problem. Consistent with the preliminary working criteria and the established goals, the team members involved in this phase of the process must determine what information will be needed and the best sources of that information. For example, they may want to know the following things:

1. What information has been published about the problem?
2. Is there an available solution to the problem?
3. If the answer to the above is yes, who is providing it?
4. What are the advantages of their solution?
5. What are the disadvantages of their solution?
6. What is the cost?
7. Is cost a significant issue?
8. What is the ratio of time spent compared to overall costs?
9. Are there legal issues to consider?
10. Are there environmental concerns to consider?

The team members can utilize many resources to assist them in their research. A good starting point may be a search using the Internet. This may provide useful material that can serve as the focus for additional research. Other reference information may come from these sources:

1. Libraries
2. Professional associations (technical and non-technical)

3. Trade journals and publications
4. Newspapers and magazines
5. Market assessment surveys
6. Government publications
7. Patent searches and listings (U.S. Patent Office: www.uspto.gov.)
8. Technical salespersons and their reference catalogs
9. Professional experts including engineers, professors, and other scientists
10. The competition's product (How is it constructed? Disassemble the competition's product and study it.)

The team will maintain detailed files, notes, pictures, sketches, and other supporting materials to assist it as it proceeds through the remaining stages of the design process. As the team discovers supporting information, the material will be added as additional reference resources.

Stage 4: Brainstorm/Generate Creative Ideas

The basic concept involved at this stage is to develop as many potential, creative solutions to the problem as possible. The more ideas generated, the greater the likelihood of identifying a feasible solution.

A primary method for generating multiple ideas to a problem is Creative Problem Solving, using the technique called Brainstorming. With this method, a large group of individuals with various backgrounds and training are brought together to attempt to solve a problem. Every idea the group spontaneously contributes is recorded. The basic premise is that no idea is deemed too wild or illogical. No preliminary judgments are made about any member's idea, and no negative comments are allowed. The goal is to develop a long list of possible alternative solutions to the problem at hand. The group leader should be able to encourage participants to suggest random thoughts and ideas. This maximizes the generation of all possible solutions.

Some students may have had the opportunity to engage in brainstorming exercises in the past. Brainstorming can be fun and highly stimulating to the creative process. For example, how many ways can you suggest to use a piece of string and a Styrofoam drinking cup? What could be created from a trash bag that contains some old magazines, tape, and a ruler? Think of ways your student organization could earn extra funds for a field trip, etc. Brainstorming sessions typically should last no longer than 20 minutes. After that much time, it's difficult to think beyond the scope of the best ideas you've already generated.

A brainstorming session could be continued on a second occasion to allow members time to consider other possible options. When the group reconvenes, members may have new ideas or perspectives for examining the problem.

At the conclusion of this stage, the group should have a long list of potential solutions. At this stage of the process, all ideas have been included for consideration. Each idea will be evaluated eventually, but all options must be kept open at this stage.

Once the project manager is satisfied that all possible solutions have been suggested, the project likely will be cleared for the next stage.

Stage 5: Analyze Potential Solutions

In the early part of the analysis stage (Phase 1), you must narrow the ideas generated in the brainstorming stage to just a few ideas which can be subjected to more sophisticated analysis techniques. This early narrowing could include the following:

- Examine the list and eliminate duplicates. As discussed earlier, do not create limited categories but eliminate repeated ideas. If two are similar, both should remain at this point.

- Allow the group to ask clarifying questions. This could help identify duplicate ideas.

- Ask the group to evaluate the ideas. The group members can vote for their top three ideas, and those that gain the most votes will be retained for more detailed analysis.

At this point, there will be a smaller number of ideas remaining. These can be analyzed using more technical and perhaps time-consuming analysis techniques (Phase 2).

Many individuals should be involved at this stage, but the engineer will be of primary importance. The analysis stage requires the engineer's time and expertise. Here, one's training in mathematics, science, and general engineering principles are applied to evaluate the potential solutions. Some of the techniques in this phase can be time consuming, but a thorough and accurate analysis is important before the project moves to the next stage. For example, if the problem involves the development of an automobile bumper that could withstand a 20-mile-per-hour crash into a fixed object barrier, several forms of analysis could be applied:

- Common sense: Do the results seem reasonable when evaluated in a simple form? Does the solution seem to make sense compared to the goal?

- Economic analysis: Are cost factors consistent with predicted outcomes? Does each of the proposed solutions satisfy the laws of thermodynamics? Newton's laws of motion? The basic principles of the resistance of a conductor, as in Ohm's law, etc.?

- Estimation: How does the performance measure up to the predicted outcomes? If the early prediction was that some of the possible bumper solutions would perform better than others, how did they perform against the estimate?

- Analysis of compatibility: Each of the possible solutions and their related mathematical and scientific principles are compared to the working criteria to determine their degree of compatibility. For example, how would each bumper solution meet the criteria of being cost effective? What would be the size, weight, and strength of each of the proposed solutions? How easy would each one be to produce?

- Computer analysis techniques: One frequently used method is finite element analysis. With this method, a device is programmed on a computer and numerically analyzed in segments. These segments are compared mathematically to other segments of the concept. In the bumper crash example, the effects of the impact could be analyzed as a head-on crash and compared to a 45-degree angle collision or a side-impact crash. As each section is analyzed, the "worst-case scenario" can be evaluated.

- Conservative Assumptions: As discussed in Chapter 6, this technique can be most useful in analysis since it can build safeguards into the analysis until more data are generated by ensuring that all minimal levels of safety, for instance are met.

After each of the working criteria have been examined and compared to the list of possible solutions, a process eliminates those that have not performed well in the various forms of analyses. At this point, three to five options from the original list of prospective solutions might remain. The project manager will review the remaining options, and he or she will authorize the project to be cleared for the next stage, assuming these remaining options meet the working criteria and overall goals. If not, the process will need to be terminated, or returned to an earlier stage to correct the problem.

Stage 6: Develop and Test Models

Once each of the prospective solutions has been analyzed and the list of feasible options has been narrowed to a few possibilities, it is time to enter the phase where specific models are developed and tested. Again, one must have a strong background in engineering coupled with experience and sound judgment. However, this stage will involve team members who are computer specialists, shop workers, testing technicians, and data analysts.

Modeling is a method of illustrating a solution to a practical problem. Students need to understand that engineers and technicians can use different forms of modeling to assess and develop a product. The types of modeling are descriptive, functional, mathematical, computer, scale, and diagram/graphical.

Descriptive modeling may include diagrams, graphs, flow charts, and block diagrams. Verbal modeling describes the behavior of systems. Mathematical modeling in the form of equations shows the relationship between variables in the systems. Scale models may be included in descriptive modeling. Examples of descriptive modeling might include a 3-dimensional CAD drawing of a structure; a verbal description of the materials to be used, heat flow formulas, and scale models of the actual systems.

Functional modeling may include computer simulations that support investigations of system behaviors or physical models of real systems with moving parts.

With mathematical models, various conditions and properties can be mathematically related as functions and compared to one another. Often these models will be computerized to assist in visualizing the changing parameters in each of the models.

Various computer models can be used. Typically these models allow the user to create on-screen images which can be analyzed prior to the construction of physical models. The

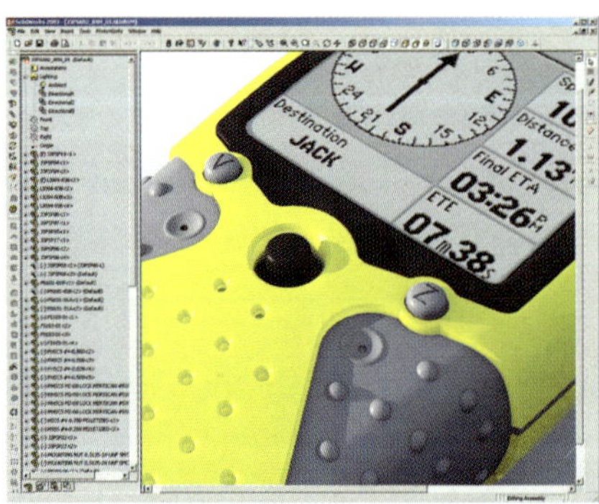

Garmin International used surfacing and geometric modeling capabilities to set new standards for innovation in hand-held GPS design. Garmin International leverages SolidWorks® integration with other applications, such as CircuitWorks for designing printed circuit boards and eDrawings for communicating with tooling developers, to accelerate time-to-market.

most common computer modeling is Computer-Aided Design (CAD) where models are designed and displayed as three-dimensional wire-frame drawings or as shaded and colored pictures. The computer can control equipment that generates solid models using techniques such as stereo-lithography, where quick-hardening liquids are shaped into models or other forms for rapid prototyping. These on-screen models, or the prototype models they produce, can be used in the testing process. Animations of solutions can effectively communicate an idea. Various software can incorporate items created in a CAD program and import them to represent them in their intended environment

With scale models, typically, these smaller models have been built to simulate the proposed design but may not include all of the particular features or functions. These models are often called prototypes or mock-ups and are useful in helping engineers visualize the actual product. Such models may be used to depict dams, highways, bridges, new parts and components, or perhaps the entire body of a prototype automobile.

In using diagrams or graphs, these models provide a tool for visualizing, on a computer or paper, the basic functions or features of a particular part or product. These diagrams or graphs could depict the electrical circuit components of an operating unit of the product or a visualization of how the components eventually may be assembled.

Zurich Unique Airport Dock E Lounge Chairs, psMetalltechnik

IDEAS Geodesic Dome

CASE STUDY

Whether societies dwell in cold or hot climates, shelter from the elements is important for survival. Often the quest for a sheltering structure may not be a need, but a want, desire, or a means to achieve some other result. Whatever the reason, a building is normally designed to shelter people and their property.

Ancient civilizations built both small and massive structures to serve a variety of purposes. Some people built temples or burial sites. Most all had the need for housing. Structures for entertainment or for the pursuit of knowledge were soon added to the list, and offered a unique twist to the typical structure. Materials used in the construction of these temples, palaces, homes, and businesses have ranged from coarse stone to marble, and from ice to sticks. Industrialization introduced metal and steel beams, and the use of glass and concrete eventually became popular.

How many different or unique structures can you think of? Consider ancient civilizations and contemplate how their structures compared to those today. Examine the dwellings of other living organisms and creatures. What do their homes look like and what factors influenced their choice of building materials?

Look at each of the structures that you or your classmates have identified and imagine each as a basic shape. Eliminate the square and rectangular types of buildings and focus on those created from a triangle. Did you know that circular structures may be made from many, smaller triangles. For instance a geodesic dome is actually comprised of triangles positioned in a specific pattern.

CHALLENGE:

EXPLORATORY

Research and model at least three (3) structures that have been constructed from triangles. Write a brief paragraph that describes their construction and why a particular culture may have selected that shape.

INTERMEDIATE

Research, design, and model a icosahedron that incorporates a lesson from another discipline. Create a game using the structure as a die or template. Use information from the other discipline to develop a set of questions (e.g., Quiz Bowl).

ADVANCED

Design and make an icosahedral structure or geodesic dome that may be used as a mini-planetarium. The structure should be large enough to accommodate six to eight students comfortably as well as the Star Machine.

MATERIALS:

EXPLORATORY

- Bristol board
- Kite string
- Heavy weight paper, various colors
- Rulers
- Sand paper, other textured paper
- Scissors
- Pencils, markers, colored pencils
- Glue

INTERMEDIATE

- Bristol board
- Kite string
- Heavy weight paper, various colors

- Rulers
- Pencils and markers
- Colored pencils
- Scissors
- Materials from other disciplines related to questions, answers, terms
- Glue

ADVANCED

- Butcher paper
- Duct tape: 6 rolls
- Meter sticks
- Kite string
- 3 Rolls of 10' x 25' 3-mm black polyethylene
- Pencils and markers
- Scissors
- Star Machine
- Box Fan

Your success on this Challenge will be based on your completion of the activities below. Three general criteria for your performance will be your participation in the activity, the accuracy of your measurement and construction of the model, and the performance of your design. Your teacher will help you understand how your performance will be graded.

EXPLORATORY

- Locate and document at least three (3) examples of structures that have been constructed from triangles.
- Write a brief paragraph that summarizes the construction techniques and why the particular culture/society may have chosen the design.
- Construct a model of the structure showing the triangles used in design.
- Present your design.

INTERMEDIATE

- Research geometric shapes based upon an equilateral triangle.
- Research icosahedrons and the construction of this shape.

- Gather review materials from other classes or disciplines. Compose questions/answers on a separate sheet of paper.
- Model an icosahedron with a height of 8 inches.
- Design a game that would utilize review materials and the icosahedral shape.
- Construct your design using the materials provided and approved.
- Test and present your design.

ADVANCED

- Research geometric shapes based upon an equilateral triangle.
- Research icosahedrons and geodesic domes and the construction of these shapes.
- Gather information about planetariums as well as astronomy facts.
- Design and create a fictitious, scientific company manufacturing a portable mini-planetarium that will travel from school to school.
- Create a brochure that will promote the new design.
- Calculate the size of the structure that will accommodate six to eight students comfortably. Remember, the Star Machine must be included in the structure.
- Construct your design using only the materials provided and approved.
- Test and present your design.

The following is an example of an integrated activity to show the possible Science, Math, and Technology (SMT) connections to the above challenges.

Using the following equation, calculate the height of the dome at different scales (e.g., full size, half size, quarter size, 1/8th size. Calculate proportions for each of the activities, whether small or large scale.

Height of Dome = 1/3 length of a triangle's side x 5
example: Height dome = 1/3x(130 cm) x 5
Height dome = 216.70 cm

A dome of the size in the example equation will accommodate six to eight students plus the Star Machine. Smaller icosahedrons may be constructed for other applications.

Method of constructing an equilateral triangle:

1. Begin with selected base line AB. (Note that AB is the length of a triangle's side in the above equation.)
2. Place compass point on A and the pencil point on B; draw a vertical arc.
3. Place compass point on B and the pencil point on A; draw a second vertical arc.
4. The point where the two arcs intersect is point C or the third corner of the Equilateral Triangle.
5. Connect points A to C and B to C.

PROBLEM SOLVING PROCESS:

These steps may be helpful to you in approaching your activity:

- Form cooperative groups (2 to 3 people).
- Brainstorm for ideas.
- Sketch possible solutions.
- Decide how to build and finish the project.
- Decide on and gather materials.
- Construct your design.
- Test your design.
- Present your design.

STUDENT REFLECTION SHEET

We all have unique ways of communicating ideas, concepts, and memories. When we record them in a logical manner using words, sentences, and pictures, we reflect upon our accomplishments as well as our failures. Quite often, we rely upon these reflections to improve or change the way we will do something. Complete the questions in your workbook using words, sentences, pictures, and stories. Be honest and record important and meaningful ideas.

IDEAS is the result of a project funded by The Engineering Foundation and organized by two of the worlds major engineering societies, the American Society of Mechanical Engineers (ASME) and the American Society of Civil Engineers (ASCE). Great lakes press has been granted permission by ASME and ASCE for the use of IDEAS projects within this textbook

MODEL TESTING

Once the models have been developed and created, it is time to test each of them. Performing various tests on each of the models allows for comparison and evaluation against the working criteria and the overall established goals. Tests are done continually throughout the stages of a project including early models, prototypes, and the testing of product quality as the product is manufactured or built. However, the results of the testing done in this stage establish the foundation for the decisions that will be made about the project's future.

Examples of these tests include the following:

Durability: How long will the product function during testing before failure? If the product is a structure, what is its predicted life span?

Ease of assembly: How easily can it be constructed? How much labor will be required? What possible ergonomic concerns are there for the person operating the equipment or assembling the product?

Reliability: These tests are developed to characterize the reliability of the product over its life cycle and to simulate long-term use by the customer.

Strength: Under what forces or loads is a failure likely, and with what frequency is it likely to occur?

Functionality: The product or solution must fulfill its intended purpose. If a design is sent to a customer and the customer cannot use the solution for its work, the design is worthless. If a mechanical organizer is designed for a parts company, it must fit in the intended space and use a power supply specified for the space in which it will be installed.

Environmental Considerations: Can the parts be recycled?

Quality and consistency: Do tests show that product quality is consistent in the various stages? Can the design be consistently manufactured and assembled? What conditions need to be controlled during manufacture or construction to ensure quality? The product or solution must be designed to meet certain minimum standards. Quality has to be in relationship to the conditions of the item's intended use. This spin-off of functionality is important because quality is defined according to the proper or improper use of the item. If casual shoes were intended to be used to play basketball, would they have quality issues when the users used them in the incorrect environment? Quality has to be evaluated in the context in which the product is intended to be used.

Safety: Is it safe for consumer use? The product must be designed to comply with codes and regulations to provide safe use by the user.

Consistency of testing: This technique examines each of the testing methodologies against the various results obtained to determine the consistency among the various tests. In the bumper design example, the testing methodology might be different for a head-on impact test than for a 45-degree crash test in order to minimize testing inconsistencies.

Ergonomics: The product must be designed so the user can operate it with ease and maximum efficiency. Ergonomics relates to human factors engineering.

All of this information will be evaluated by the project manager. If the manager is satisfied these results consistently meet the working criteria and overall goal, the project likely will be cleared for the next stage.

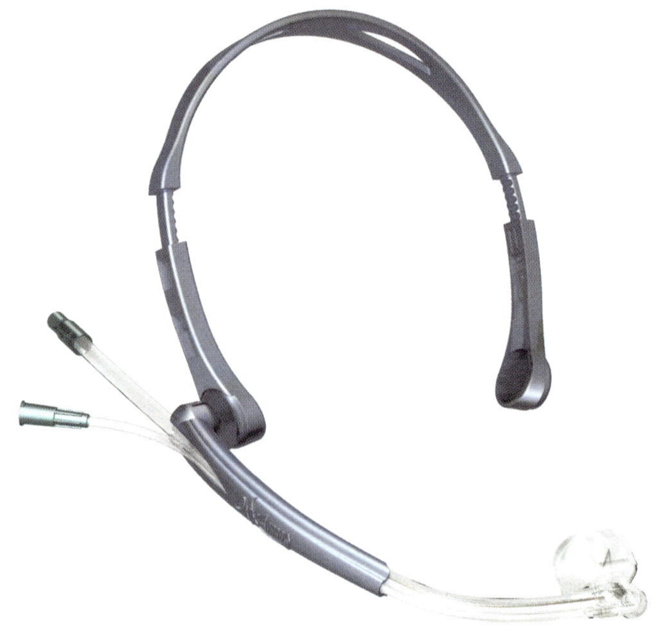

OxyArm™ Predictable Oxygen Delivery System, SOUTHMEDIC, INC.

Perishable Fruit Container

CASE STUDY

Fruit and vegetables are shipped in containers everyday. These containers may come from across town or from other cities or countries. Your team's objective is to design and construct a container that will hold 25 pieces of fruit. The fruit is from a tropical rainforest in Africa. Because this fruit grows only in a remote area, all supplies must be air lifted in and out. Due to the limited space, your package must be sent to Africa flat and then be assembled by the villagers, so your team must supply directions for them. Because they do not understand English, your instructions must not include any numbers or letters.

The fruit is sensitive to temperature change so your package must protect the fruit from sudden or gradual temperature change. It must also protect it from being damaged during shipment.

This case study will involve many problem-solving methods. You must work as a team and assign members specific jobs to do. The rules for the container development are listed below. These must be followed by every team. These rules reflect a Request For Proposal (RFP), which is how companies, the government, and other agencies outline their requirements for a project.

According to Lilley Information Systems, "An RFP in its most formal sense is a specification of requirements that is sent out to suppliers who reply with proposals. Although common with large companies, the idea can be applied usefully at varying levels of sophistication to small - and medium - sized organizations as well. Used properly, it is a tool that supports and protects the buyer."

RFP

TO: Engineering Team
SUBJECT: Request For Proposal
PROJECT: Shipping container, perishable fruit design
AGENCY: Fruit Broker, USA

PROJECT DESCRIPTION: Design a container that will hold 25 pieces of fruit. The container is to be shipped flat and assembled on site with universal directions included. The container must protect the fruit from normal shipping impacts.

1. Container must hold 25 pieces of fruit. The fruit is the size of a tennis ball.
2. The container may be made from any material. Standard cardboard and 12 inches of clear packaging tape will be provided. Other materials will be at your team's expense. DUCT TAPE IS NOT ALLOWED.
3. The instructions for assemble of the package must be in pictograms or cartoon format so language is not an issue.
4. As space and cost are important you will receive bonus points for using less material. Your team will receive bonus points for how easy the container can be packed (see assessment).
5. Since the fruit is fragile, your package must maintain a relatively constant temperature. A 10-degree temperature drop is acceptable. One point per degree beyond 10 degrees will be deducted.
6. 12 inches of packaging tape is allowed for assembly only. Five bonus points will be awarded if the tape is not used.
7. The container must withstand a four-foot drop

Stage 7: Make the Decision

At this stage, team members must establish a means to compare and evaluate the results from the testing stage to determine which, if any, of the possible solutions will be implemented. The working criteria that have been used throughout the process are the critical factors used to determine the advantages and disadvantages of each of the remaining potential solutions. One way to evaluate the advantages and disadvantages of each of the proposed solutions is to develop a decision table to help the team visualize the merits of each. Typically, a decision table lists the working criteria in one column. A second column assigns a weighted available point total for each of the criteria. The team will need to determine the order of priority of each of the criteria. The third column then provides performance scores for each of the possible solutions. A sample Decision Table might look like the one that follows.

TABLE 16.1 A Decision Table

Working Criteria	Points Available	#1	#2	#3
Cost	20	10	15	18
Production Difficulty	15	8	12	14
Size, Weight, Strength	5	5	4	4
Appearance	10	7	6	8
Convenient to use	5	3	4	4
Safety	10	8	7	8
Legal issues	5	4	4	4
Reliability/Durability	15	7	9	11
Recyclability	5	4	3	4
Customer Appeal	10	7	8	9
Total	**100**	**63**	**72**	**84**

Based on this sample decision table, though none of the proposed solutions have scored near the "ideal" model, Solution #3 did perform better than the others. Using this information, the project manager and team leaders would make the final "go" or "no-go" decision for this project. They may decide either to pursue Solution #3, to begin a new process, or even to scrap the entire project. Assuming they decide to pursue Solution #3, the team would prepare the appropriate information for the next stage in the process.

Stage 8: Communicate and Specify

Before a part, product, or structure can be manufactured, there must be complete and thorough communication, reporting, and specification for all aspects of the item. Engineers, skilled craft workers, computer designers, production personnel, and other key individuals associated with the project must work together at this stage to develop the appropriate written materials. Such materials include detailed written reports, summaries of technical presentations and memos, relevant e-mails, diagrams, drawings and sketches, computer printouts, charts, graphs, and any other relevant and documented material. This information will be critical for those who are involved in granting final project approval and for the group involved in the final product implementation. They must have complete knowledge of all parts, processes, materials, facilities, components, equipment, machinery, and systems that will be involved in production.

Communication is an important element throughout the design process but is especially important at this stage. If team members cannot adequately sell their ideas to the rest of the organization and describe the precise details and qualities of the product or process, then many good possible solutions might be ignored. At this stage, training materials, operating manuals, computer programs, or other relevant resources may have to be created for the sales team, legal staff, prospective clients, and customers to use.

If the project manager is satisfied that all necessary materials have been adequately prepared and presented, the project likely will be passed on to the next stage, the implementation stage.

Stage 9: Implement and Commercialize

The next-to-last stage of the design process represents the final opportunity for project revision or termination. At this point, costs begin to escalate dramatically, so all serious issues should be resolved by this time.

In addition to the project manager and team leaders, other individuals representing various backgrounds and areas of expertise will be involved at this stage. Though engineers are involved at this stage, many of the activities here are performed by others. Some of those involved in this stage may include the following:

1. Management and key supervisory personnel. These individuals will make the ultimate decisions concerning the proposed project. They are concerned with monitoring the organization's long-term goals and objectives, determining future policies and programs that support these goals, and making the economic and personnel decisions that affect the overall health of the organization.

2. Technical representatives. These may include skilled craft workers, technicians, drafters, computer designers, machine operators, and others involved in manufacturing and production. This group will have primary responsibility for getting the product "out the door."

3. Business representatives. This group may consist of the following:
   a. Human resource personnel if new individuals must be hired
   b. Financial people to handle final budget details and financial analysis questions
   c. Purchasing personnel who will procure the needed materials and supplies
   d. Marketing and advertising staff members who will help promote the product
   e. Salespeople who will be involved in the actual selling and distribution of the product

4. Attorneys and legal support staff. These are the legal representatives who will handle the legal issues including patent applications, insurance, and risk protection analysis.

If all parties are in agreement that all criteria have been satisfied and the overall goal achieved, the actual production and commercialization will begin. However, one stage remains where the project activities and processes will be evaluated and reviewed. (Note: Some stages of the process may be monitored differently depending on whether the project relates to a one-time product - i.e., a bridge or a dam - or an ongoing manufacturing process.)

Stage 10: Perform Post-Implementation Review and Assessment

At this point, it is assumed that the project is in full production. The project manager, key supervisory personnel, and team members who had significant input on the project are gathered together for a final project review and assessment. This stage involves the termination of the project team since the product is now considered to be a regular product offering in the firm's overall product line. The product's performance is reviewed, including the latest data on production efficiency, quality control reports, sales, revenues, costs, expenditures, and profits. An assessment report is prepared which will detail the product's strengths and weaknesses, outline what has been learned from the overall process, and suggest ways the teams can improve the process quality. This report will be used as reference for future project managers and teams to consult.

Future City Competition

CASE STUDY

Cities of the future must maintain the quality of life for their citizens and the efficient use of land within their borders. To accomplish this, city governments will increasingly rely on redevelopment projects to improve vacant land or to rehabilitate existing structures. Prime locations for redevelopment include downtowns, transit corridors, and locations near employment, shopping, and recreational centers, and cultural amenities. City governments will actively promote redevelopment of areas for the following reasons:

1. Redevelopment reuses properties that may have been underutilized or blighted, helping to encourage revitalization.
2. Redevelopment has the potential to boost jobs, purchasing power, and public amenities in neighborhoods, and to generate tax dollars for local government.
3. Redevelopment projects can reduce pollution.

THE CHALLENGE: DESIGN AND BUILD A FUTURE CITY

Design a future city using SimCity 3000™ competition software. Teams will design and build a city of the future. The city must display residential, commercial, and industrial areas, power plants, roads, and power distribution networks. The city must be energy -efficient, supplying enough energy for its residents. Other considerations are pollution levels, traffic density, and cost efficiency. The solution includes developing a project plan, a logical model of the city using SimCity3000™ software, a team-made physical scale model, an essay, an abstract, and a verbal presentation.

FUTURE CITY COMPETITION OVERVIEW

PHASE I

Computer City Design (disk)

With the help of the engineer mentor and teacher-sponsor, design a logical model of the city using SimCity3000™ software.

PHASE II

Build Model

Your team will construct a physical model of the city using recycled materials. The model can be no larger than 30" (W) x 60" (L) x 24" (H).

PHASE III

Essay

Write a 500-word to 700-word essay on “Creating an Engineering Feasibility Plan” for a specific redevelopment area in a Future City. They also will write a 300-word to 500-word abstract describing their city and some of its services.

PHASE IV

Verbal Presentation

Explain the unique design features of your city during the team verbal presentation to a panel of judges.

Chapter 8

Engineering Fundamentals and Resources

This chapter introduces basic physical principles and concepts behind mechanisms related to the fields of engineering and technology. Some of the math may be a bit advanced for your particular course, but the material will be helpful to you at some point in your studies. In some classes, only the concepts presented here will be discussed while the math will be ignored. The important thing is for you to expose yourself to the concepts and the "language" of engineering and technology. For more detailed discussions of many of these topics, please visit the following URL: http://www.glpbooks.com/hs/reference.

Several conventions should be reviewed at the outset in case you are not familiar with them or have forgotten them from a previous class, though often they will be discussed in greater detail throughout the chapter. Many figures and diagrams will present arrows that are intended to show the direction of forces applied. Fixed surfaces will be presented as a straight line with shading or a series of angled lines to one side. Some of the equations presented in this chapter will enlist a small triangle () in front of variables to signify a change in that variable, such as a change in the location of an object (distance traveled) or a change in (elapsed) time. Feel free to consult your instructor regarding any questions you may have as you work through this chapter.

1 PHYSICAL PRINCIPLES

1.1 Force and Torque

1.1.1 Force

Force: *An agent or influence that, if applied to an object, results mainly in an acceleration of the object.*

Every day we deal with *forces* of one kind or another. A pressure is a force. The earth exerts a force of attraction (gravity) for all bodies (objects, etc.) at or near its surface. To study the forces acting on objects, we must know how the forces are applied, the direction of the forces, and their value. Graphically, forces often are represented by a *vector*, depicted by an arrow whose pointed end represents the direction of action. Vectors are specified by both a magnitude and a direction. This will be covered in more detail in the section on Momentum, Work, and Energy.

In engineering, a *mechanism* is what is responsible for any action or reaction. *Machines* are based on the idea of transmitting forces through a series of predetermined motions. These related concepts are the basis of dynamic movement.

1.1.2 Torque

Torque: *Something that produces or tends to produce rotation, and whose effectiveness is measured by multiplying the force times the perpendicular distance from the line of action of the force to the axis of rotation.*

1.2 Motion

Motion*: A change of position or orientation.*

1.3 Newton's Laws of Motion

1.3.1 Newton's First Law of Motion

When no force is exerted on a body, it stays at rest or moves in a straight line with constant speed. This principle of *inertia* is also known as ***Newton's First Law of Motion.*** It is from this law that Sir Isaac Newton was able to develop our present understanding of dynamics.

1.3.2 Newton's Second Law of Motion

From our daily life, we can observe that:

1. When a force $\mathbf{F}$ is applied on an object, then $\Delta\mathbf{V}$, the change of the velocity of the object, increases as time increases, Δt;

2. The greater the force **F** applied to an object, the greater the $\Delta\mathbf{V}$; and
3. The larger the object is, the less easily it is accelerated by a force **F**.

1.5 Work, Power, and Energy

1.5.1 Work

Work is a force applied over a distance. If you drag an object along the floor you do work in overcoming the friction between the object and the floor. In lifting an object you do work against gravity, which tends to pull the object toward the earth. Steam in a locomotive cylinder does work when it expands and moves the piston against the resisting forces. Work is the product of the resistance overcome and the distance through which it is overcome.

1.5.2 Power

Power is the rate at which work is done.

In the British system of units (which uses the terms feet, pounds, etc.), power is expressed in foot-pounds per second. For larger measurements, horsepower, hp, is used.

$$1\text{hp} = 550 \text{ ft*lb/s} = 33{,}000 \text{ ft*lb/min}$$

In SI units (which uses the terms meters, grams, etc.), power is measured in joules per second, also called the *watt* (W).

$$1\text{hp} = 746 \text{ W} = 0.746\text{kW}$$

1.5.3 Energy

All object possess *energy*. Generally, there are two kinds of energy in mechanical systems: *potential energy* and *kinetic energy*. Potential energy is due to the position of the object and kinetic energy is due to its movement.

For example, an object set in motion can overcome a certain amount of resistance before being brought to rest, and the energy that the object has is due to its motion. *Flywheels* on engines both receive and give up energy related to motion (kinetic energy).

Elevated weights have potential to do work on account of their elevated position, as is the case with a raised hammer, for instance.

2 MECHANISMS AND SIMPLE MACHINES

The study of machines and mechanisms has always been vital to human progress. In the first century BC, the Roman author Vitruvius identified five simple machines for moving loads. These were the lever, the pulley, the windlass, the wedge (or inclined plane), and the screw. They were all much in use at that time, especially in building some of the magnificent monuments of that age, and until the nineteenth century were considered to be the basic elements in all machines.

In Berlin in the late nineteenth century, Franz Reuleaux defined and developed the fundamental concepts of *kinematics*. He listed six mechanical elements on which all machines and mechanisms were founded. Four of them were the lever (or crank), the wheel and gears, the cam, and the screw.

He identified another element that was limited to transmitting either tension or compression, examples being pulley ropes, transmission belts or chains, and hydraulic fluid lines. The sixth fundamental element of kinematics that he identified was that which transmitted intermittent motion, such as a ratchet.

A *machine* is a combination of moving (and fixed) mechanical components that transmits or modifies the action of a force or torque to do useful work. Machines such as a food mixer, food processor, washing machine, electric sander, or drill are familiar to us. An *engine* is a machine that converts a natural energy source to the output power that is required for a particular application.

Whereas a machine is primarily concerned with the transmission of forces, a mechanism is principally concerned with movement. A *mechanism* is a system that transforms one kind of motion to another. It may consist of a single component or a combination of the fundamental components: lever, cam, screw pulley, gear, and ratchet. A sewing machine and a clock are everyday examples of mechanisms.

Mechanisms can be identified in all machines; the piston-crank mechanism in an engine, and the quick return mechanism of a shaping machine, for example. It is the function of these mechanisms to provide correct positioning of the machines components at each stage in the sequence of its operation.

Again, here are some definitions of terms:

***Mechanism**: the fundamental physical or chemical processes involved in or responsible for an action, reaction, or other natural phenomenon.*

***Machine**: an assemblage of parts that transmit forces, motion, and energy in a predetermined manner.*

***Simple Machine**: any of various elementary mechanisms having the elements of which all machines are composed. Included in this category are the lever, wheel and axle, pulley, inclined plane, wedge, and screw.*

The word *mechanism* has many meanings. In *kinematics*, a mechanism is a means of transmitting, controlling, or constraining relative movement (Hunt 78). Movements that are electrically, magnetically, pneumatically operated are excluded from the concept of mechanism. The central theme for mechanisms is rigid bodies connected together by joints.

A *machine* is a combination of rigid or resistant bodies, formed and connected do that they move with definite relative motions and transmit force from the source of power to the resistance to be overcome. A machine has two functions: transmitting definite relative motion and transmitting force. These functions require strength and rigidity to transmit the forces.

The term *mechanism* is applied to the combination of geometrical bodies that constitute a machine or part of a machine. A *mechanism* may therefore be defined as a combination of rigid or resistant bodies, formed and connected so that they move with definite relative motions with respect to one another (Ham *et al.* 58).

Although a truly *rigid body* (object) does not exist, many engineering components are rigid because their deformations and distortions are negligible in comparison with their relative movements.

The *similarity* between machines and mechanisms is that

- they both are combinations of rigid bodies
- the relative motion among the rigid bodies is definite.

The *difference* between *machine* and *mechanism* is that machines transform energy to do work, while mechanisms do not necessarily perform this function. The term *machinery* generally means machines and mechanisms. Figure 2.1 shows a picture of the main part of a diesel engine. The mechanism of its cylinder-link-crank parts is a *slider-crank mechanism*, as shown in Figure 2.2.

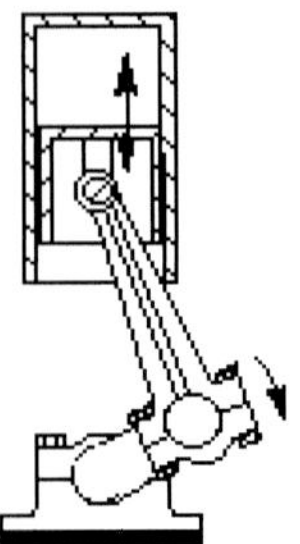

Figure 2.1 Cross section of a power cylinder in a diesel engine

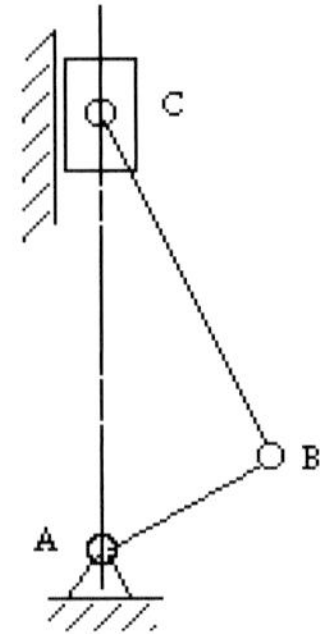

Figure 2.2 Skeleton outline

2.1 The Inclined Plane

Figure 2.3a shows an *inclined plane*: AB is the base, BC is the height, and AC is the *inclined plane*. With the use of the inclined plane, a given resistance can be overcome with a smaller force than if the plane is not used (over a flat or horizontal surface, for instance). For example, in Figure 2.3b, suppose we wish to raise a weight of 1000 lb through the vertical distance BC = 2 ft. If this weight were raised vertically and without the use of the inclined plane, the 1000-lb force would have to be exerted through the distance BC. If, however, the inclined plane is used and the weight is moved over its inclined plane AC, a force of only 2/3 of 1000 lb, or 667 lb, is necessary, though this force is exerted through a distance AC, which is greater than distance BC.

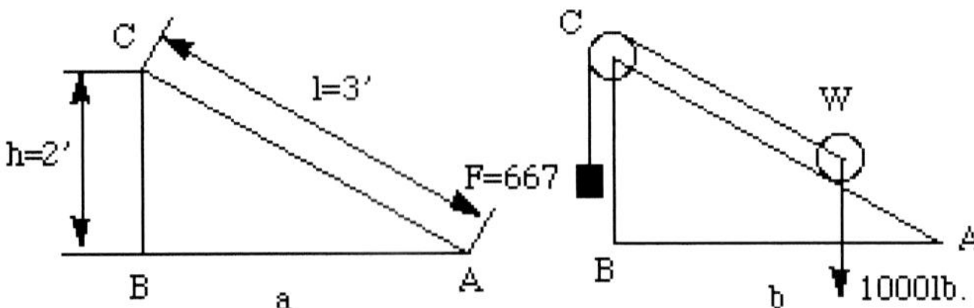

Figure 2.3 Inclined plane

Using an inclined plane requires a smaller force exerted through a greater distance to do a certain amount of work.

Letting **F** represent the force required to raise a given weight on the inclined plane, and W the weight to be raised, we have the proportion:

(2-1)

2.1.1 Screw Jack

One of the most common applications of the principle involved with the inclined plane is involved in the *screw jack*. A screw jack is used to overcome a heavy pressure or raise a heavy weight of *W* by a much smaller force *F* applied at the handle. *R* represents the length of the handle and *P* the *pitch* of the screw, or the distance advanced in one complete turn of the jack handle.

The screw jack consists of a screw running inside a nut. When an effort, applied to the handle at a distance *R* from the center of the screw, is moved through one complete turn the screw thread will advance by the *lead* distance *P*. For a single start thread, this will be equivalent to the pitch of the thread, the distance between two adjacent crowns of screw thread. For a multiple start thread the lead will be equal to the pitch multiplied by the number of starts.

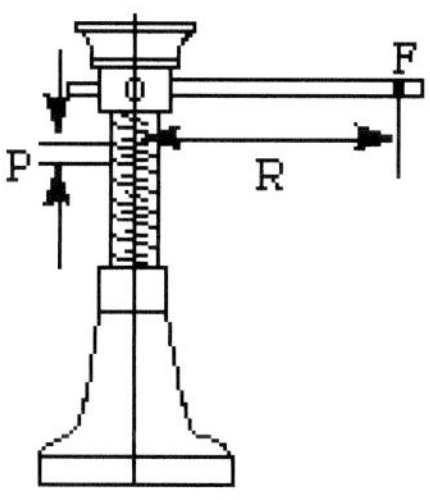

Figure 2.4 The screw jack

2.2 Gears

A gear, or toothed wheel, when in operation, may actually be considered as a lever with the additional feature that it can be rotated continuously, instead of rocking back and forth through a short distance. Essential elements related to a gear are the number of teeth, the diameter, and the rotary velocity of the gears. Figure 2.5 shows the ends of two shafts A and B connected by two gears of 24 and 48 teeth, respectively. Notice that the larger gear will make only one-half turn during the time the smaller gear makes a complete turn. That is, the ratio of speeds (velocity ratio) of the large to the smaller is 1 to 2.

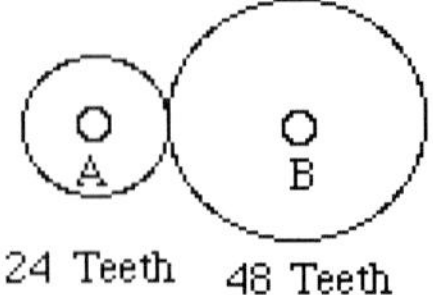

Figure 2.5 Gears

The gear that is closer to the source of power is called the *driver*, and the gear that receives power from the driver is called the *driven gear*.

2.2.1 Gear Trains

A gear train may have several drivers and several driven gears. Figure 2.6 shows two meshed spur gears of different sizes. This is an example of a simple gear train, with one gear on each shaft. The smaller, 9-toothed pinion gear will have to perform two revolutions for each revolution of the larger, 18-toothed wheel gear, so when the wheel is used as the driver or input gear, the output motion will be faster than the input

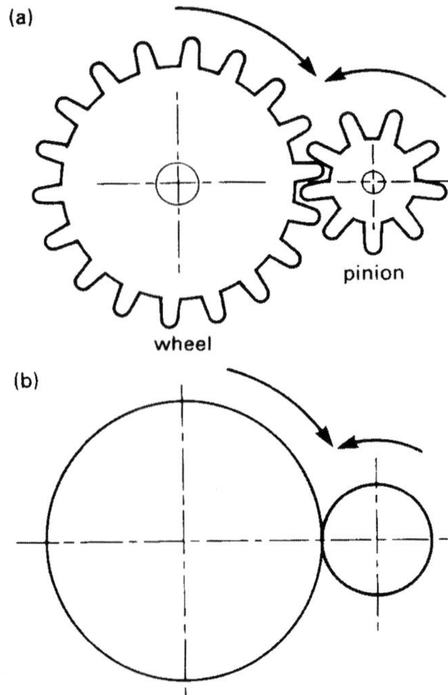

Figure 2.6 (a) & (b) ***Simple gear train showing the direction of rotation of the meshed spur gears.***

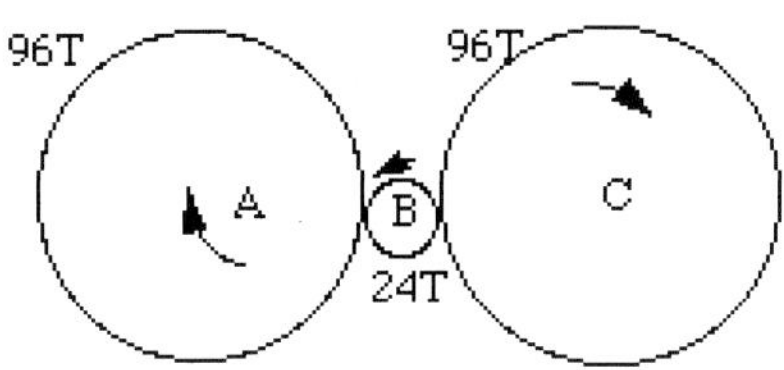

Figure 2.7 Gear train

In Figure 2.7 above, when gear A turns once clockwise, gear B turns 4 times counter-clockwise, and gear C turns once clockwise. Hence gear B does not change the speed of C from what it would have been if geared directly to gear A, but it changes its direction from counterclockwise to clockwise.

The velocity ratio of the first and last gears in a train of simple gears does not change by putting any number of gears between them.

2.3 Chains, Belts and Pulleys

2.3.1 Chains and Belts

Apart from gears there are two main methods of transmitting power between two shafts: *chain drives* and *belt drives.* Belts are cheaper, and gears more costly to manufacture, assemble, and align. The choice of the type of transmission depends mostly on four factors:

- The distance between the shafts
- The required performance
- The cost
- The maintenance of the system

If the center distance is large, a belt drive is appropriate, but over small distances gears are more practical. If precise performance and accurate timing over a large distance are required, a flat belt, which will tend to slip, will not prove satisfactory, and a more expensive means must be employed such as a chain or toothed-belt drive. These factors explain the use of chain, gears, or toothed belt to drive the camshaft of a car, but usually a *V-belt* is used to increase the friction between the belt and pulley to drive the fan and alternator in a vehicle.

Belt drives were once commonplace in industrial machinery, but it is now more common for the distribution of power to occur though electrical cables when powering individual motors. Belt drives are still used in many household applications such as washing machines and computer printers. In many cases a stepper motor is used in conjunction with the belt transmission to achieve a smoother drive.

2.3.2 Pulleys

Pulleys are nothing but gears without teeth. Instead of running together directly, they are made to drive one another through the use of cords, ropes, cables, or belting of some kind.

As with gears, the velocities of pulleys are inversely proportional to their diameters (i.e., they rotate faster the smaller they are).

Figure 2.9 Belts and pulleys

From the earliest times, people have needed to raise weights grater then they could lift on their own. Consequently, they developed lifting machines with a large mechanical advantage to lift heavy weights more easily. The simplest of these is based on the lever, such as a crowbar used for prying objects apart.

2.3.3 Simple pulley systems

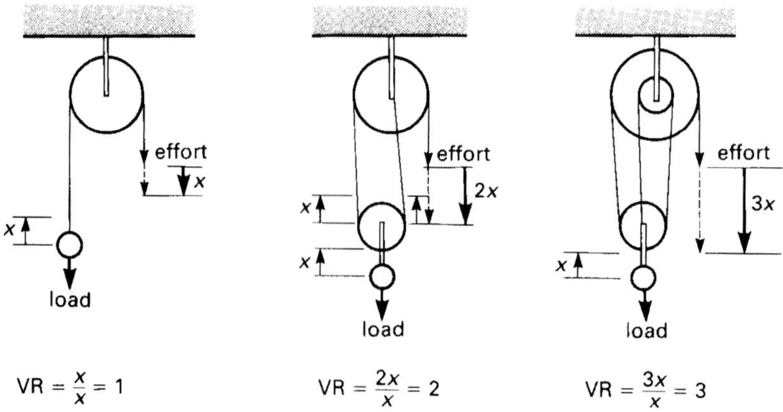

Figure 2.11 Pulley lifting devices

Figure 2.11 shows three pulley systems, each with a continuous rope wound around the pulley sheaves giving 1, 2, or 3 sections of rope supporting the load. Pulley systems such as these are often used as lifting machines and can have a very high efficiency. The velocity ratio of each system can be found by considering the movement of the effort and the load, and it can be seen from figure 2.11 that the velocity ratio is equivalent to the number of rope sections supporting the load.

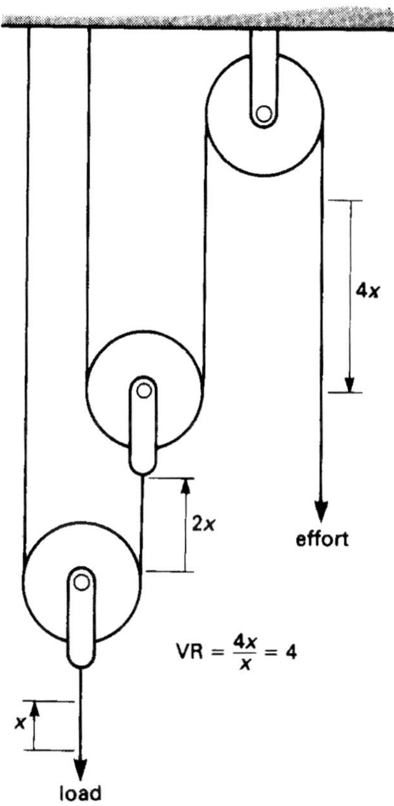

Figure 2.12 Pulley lifting system

The arrangement of pulleys in figure 2.12 is different from that in the previous example, but by again examining the effort distance for a set load distance, we find that the velocity ratio of this system is 4.

2.4 Lever

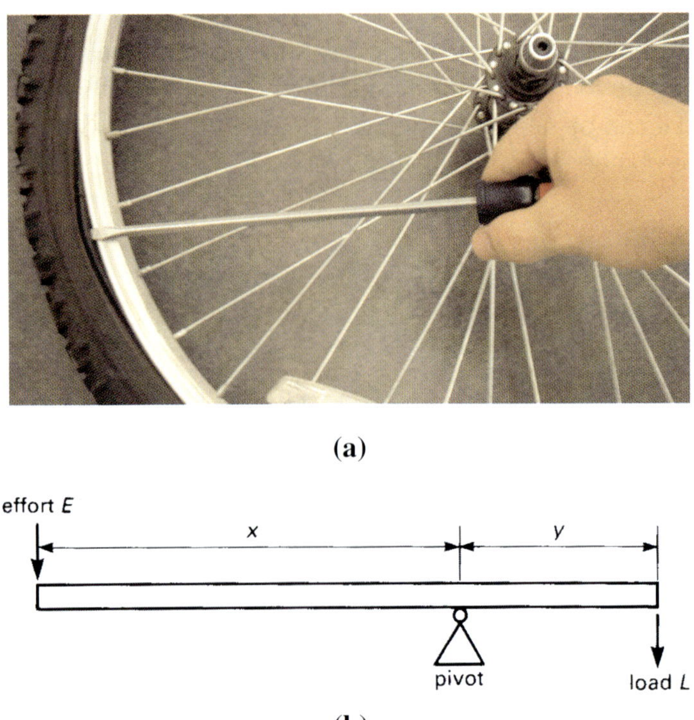

Figure 2.13 A class 1 lever
(a) A lever removing a bike tire (b) The forces in a class 1 lever

Figure 2.13(a) shows a lever being used to force a bicycle tire from its rim. This can be considered to be a rigid *lever* pivoted at the fulcrum. The effort applied at one end lifts a load placed at the other, as shown in fig 2.13(b). The bar and also the scissors in figure 2.14 are examples of *class 1 levers* in which the pivot or *fulcrum* is between the load and the effort. The lever will magnify a force if the load point is closer to the fulcrum than the effort. When using scissors it is noticeable that the greatest cutting force is obtained when the material to be cut is placed as near to the pivot as possible.

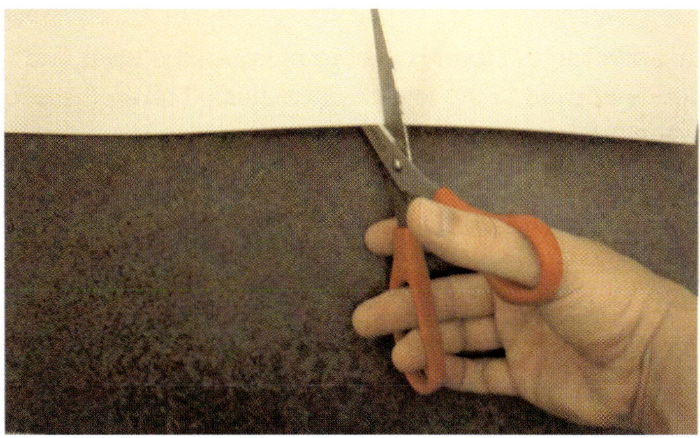

Figure 2.14 Scissors are a double-lever mechanism

Mechanical advantage (MA) refers to the way a lever or a pulley magnifies the effort or force applied. It is defined as:

$$\text{Mechanical advantage} = \frac{\text{load}}{\text{effort}} \text{ or, using variables, MA} = \frac{L}{E}$$

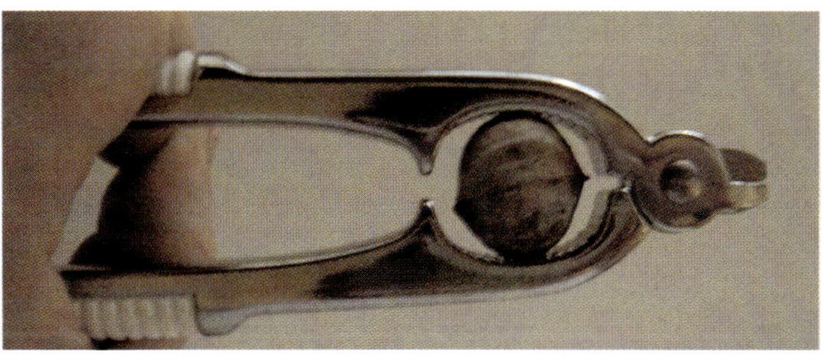

(a)

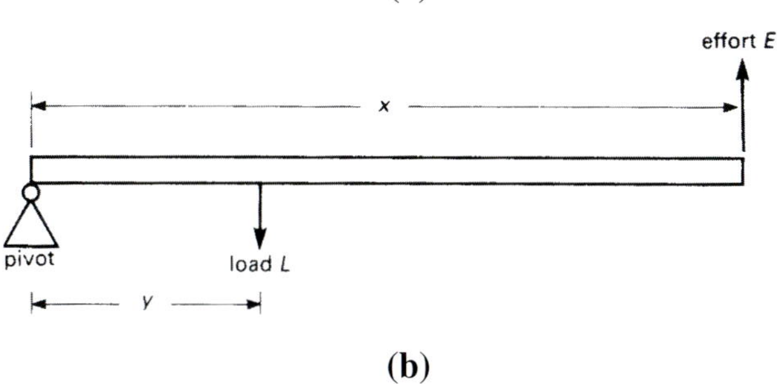

(b)

Figure 2.15 A class 2 lever
(a) A lever used to crack nuts (b) The forces in a class 2 lever

With a class 2 lever, the load is placed between the pivot and the point where effort is applied to the lever. With a class 3 lever, the effort is applied between the load and the pivot.

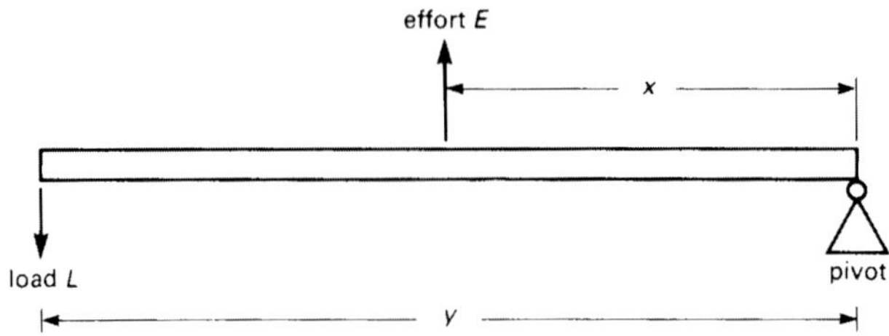

Figure 2.16 A class 3 lever

2.5 Wheel and Axle

The wheel and axle shown in figure 2.17 is a development of the early windlass, which was used to lift heavy weights by winding a rope around a pole.

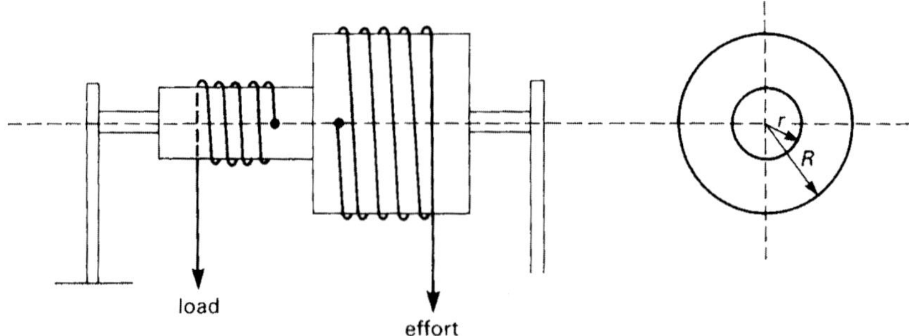

Figure 2.17 A wheel and axle

The effort wheel is of greater radius than the load drum, and both are mounted on the same axle. When effort is applied to the wheel, a mechanical advantage is obtained, and for one turn of the effort wheel the load drum will also rotate by one revolution. Hence:

$$VR = \frac{\text{Distance moved by effort}}{\text{Distance moved by load}} = \frac{2pR}{2pr} = \frac{R}{r}$$

This is also the velocity ratio for an ungeared winch in which a load drum of radius r is rotated by means of a crank of length R.

When a gear train is incorporated into the wheel and axle system then the mechanical advantage can be greatly increased. The velocity ration becomes:

$$VR = \frac{RG}{r}$$

where *R* is the effort wheel radius
r is the load drum radius
G is the gear ratio when the drive wheel is connected to the effort drum

2.6 Wedge

See inclined plane

2.7 Efficiency of Machines

In working out problems pertaining to *levers*, *belts and pulleys*, *inclined planes,* and so forth, we have not taken account of friction or other sources of energy loss. In other words, we have supposed them to be perfectly efficient when in fact they are not. To measure the performance of a machine, we often find its ***efficiency***, which is defined as

$$\eta = \frac{W_{out}}{W_{in}}$$

(2-4)

where
= the efficiency of a machine,
W_{in} = the input work to a machine, and
W_{out} = the output work of a machine.

3 MORE ON MACHINES AND MECHANISMS

3.1 Planar and Spatial Mechanisms

Mechanisms can be divided into ***planar mechanisms*** and ***spatial mechanisms***, according to the relative motion of the rigid bodies. In planar mechanisms, all of the relative motions of the rigid bodies are in one plane or in parallel planes. If there is any relative motion that is not in the same plane or in parallel planes, the mechanism is referred to as a ***spatial mechanism***. In other words, *planar mechanisms* are essentially two-dimensional while *spatial mechanisms* are three-dimensional. This chapter covers only planar mechanisms.

3.2 Kinematics and Dynamics of Mechanisms

The ***kinematics*** of mechanisms is concerned with the motion of the parts, without considering how the influencing factors (force and mass) affect the motion. Therefore, kinematics deals with the fundamental concepts of space and time and the derived quantities velocity and acceleration.

Kinetics deals with action of forces on bodies. This is where the effects of gravity come into play.

Dynamics is the combination of *kinematics* and *kinetics*. Dynamics of mechanisms is concerned with the forces that act on the parts—both balanced and unbalanced forces—taking into account the masses and accelerations of the parts as well as the external forces.

3.3 Links, Frames and Kinematic Chains

A ***link*** is defined as a rigid body having two or more pairing elements, which connect it to other bodies for the purpose of transmitting force or motion (Ham *et al.* 58).

In every machine, at least one link either occupies a fixed position relative to the earth or carries the machine as a whole along with it during motion. This link is the ***frame*** of the machine and is called the *fixed link*.

The combination of links and pairs without a fixed link is not a mechanism but a ***kinematic chain.***

3.4 Skeleton Outline

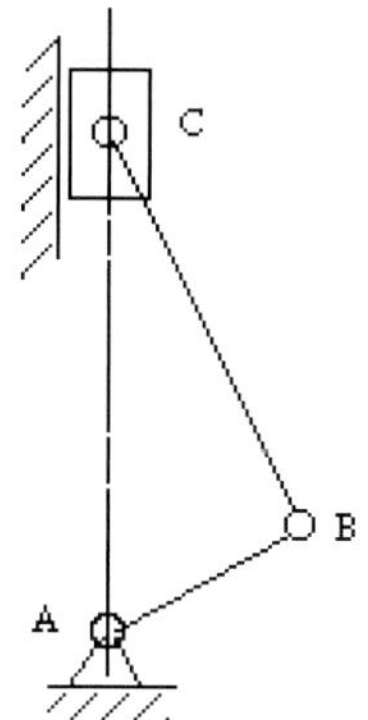

Figure 3.1 Skeleton outline

For the purpose of kinematic analysis, a mechanism may be represented in an abbreviated, or skeleton, form called the *skeleton outline* of the mechanism. In Figure 3.1, the skeleton outline has been drawn for the engine shown in Figure 2.1. This skeleton contains all necessary information to determine the relative motions of the main links—namely, the length AB of the crank, the length BC of the connecting rod, A the location of the axis of the main bearing, and the path AC of point C, which represents the wrist-pin axis. Fixed positions such as walls, floors, the ground, etc. are represented by a flat line with shading or several diagonal lines next to it.

3.5 Pairs, Higher Pairs, Lower Pairs and Linkages

A ***pair*** is a joint between the surfaces of two rigid bodies that keeps them in contact and relatively movable. For example, in Figure 3.2 a door jointed to the frame with hinges makes a ***revolute joint*** (***pin joint***), allowing the door to be turned around its axis. Figure 3-2b and c show skeletons of a revolute joint. Figure 3-2b is used when both links joined by the pair can turn. Figure 3-2c is used when one of the links jointed by the pair is the frame.

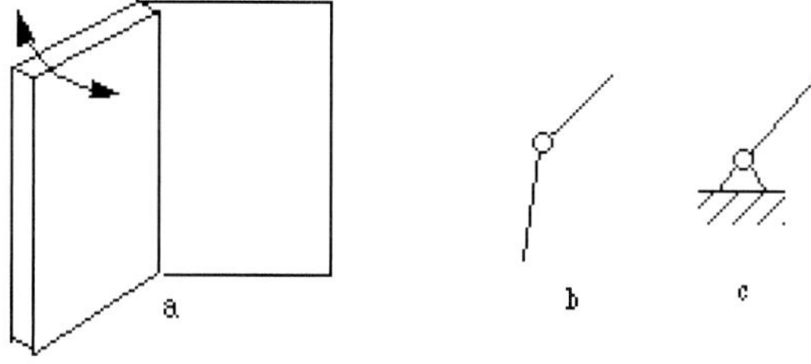

Figure 3.2 Revolute pair

In Figure 3.3a a sash window can be translated (moved) relative to the sash. When this kind of relative motion is involved it is called a ***prismatic pair***. Its skeleton outlines are shown in Fig. 3.3*b*, *c,* and *d*. Fig. 3.3*c* and *d* are used when one of the links is the *frame*.

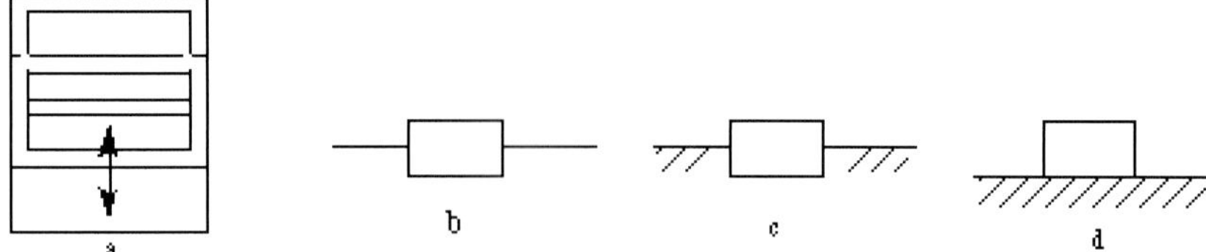

Figure 3.3 Prismatic pair

Generally, there are two kinds of *pairs* in mechanisms, ***lower pairs*** and ***higher pairs***. What differentiates them is the type of contact between the two bodies of the pair. Surface-contact pairs are called ***lower pairs***. In planar (2D) mechanisms, there are two subcategories of lower pairs—*revolute pairs* and *prismatic pairs*, as shown in Figures 3.2 and 3.3, respectively. Point-, line-, or curve-contact pairs are called ***higher pairs***. Figure 3.4 shows some examples of *higher pairs.* Mechanisms composed of rigid bodies and lower pairs are called ***linkages***.

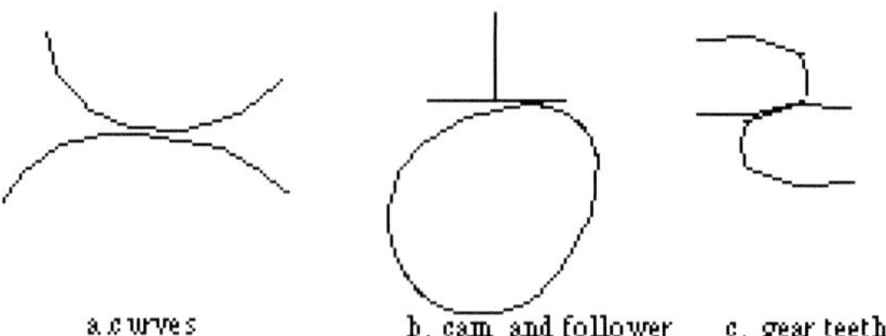

Figure 3.4 Higher pairs

4 GEARS

Gears are machine elements that transmit motion by means of successively engaging teeth. The gear teeth act like small levers.

4.1 Gear Classification

Gears may be classified according to the relative position of the axes of revolution. The axes may be:

1. parallel
2. intersecting
3. neither parallel nor intersecting

Here is a brief list of the common forms. We will discuss each in more detail later.

- gears for connecting parallel shafts
- gears for connecting intersecting shafts
- neither parallel nor intersecting shafts

Gears for connecting parallel shafts

1. Spur gears

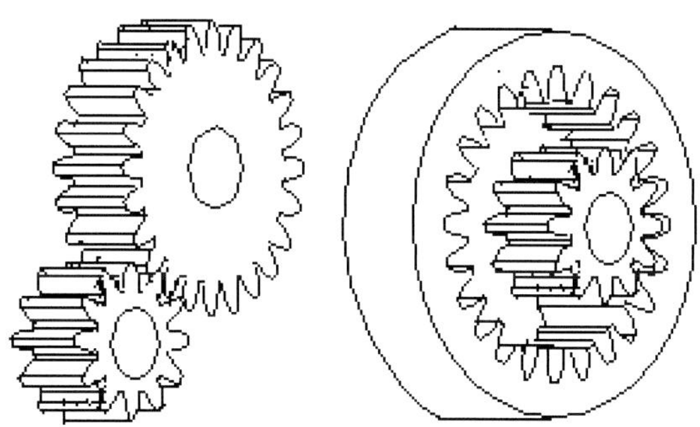

The left pair of gears makes *external contact*, and the right pair of gears makes *internal contact*.

2. Parallel helical gears

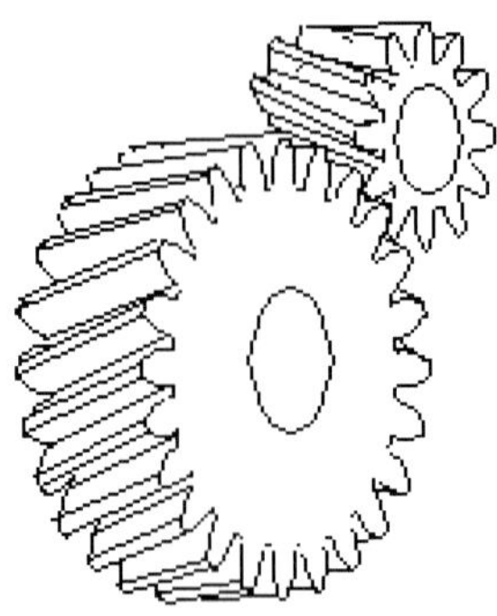

3. Herringbone gears (or double-helical gears)

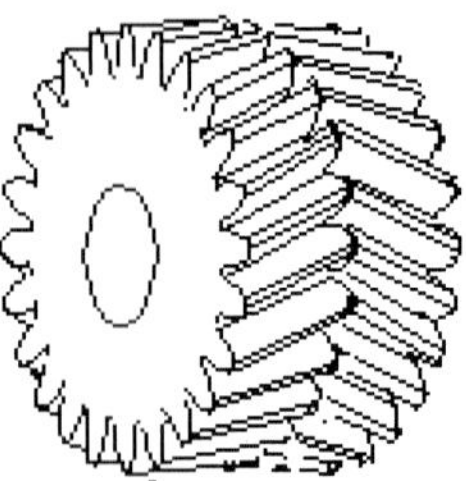

4. Rack and *pinion* (The rack is like a gear whose axis is at infinity.)

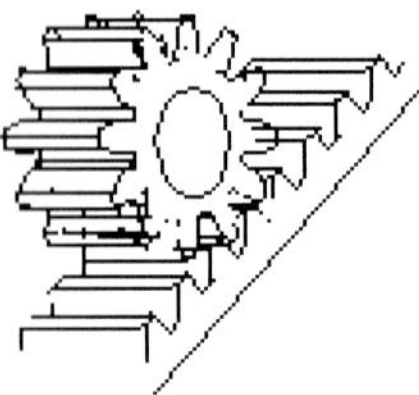

Gears for connecting intersecting shafts

Straight bevel gears

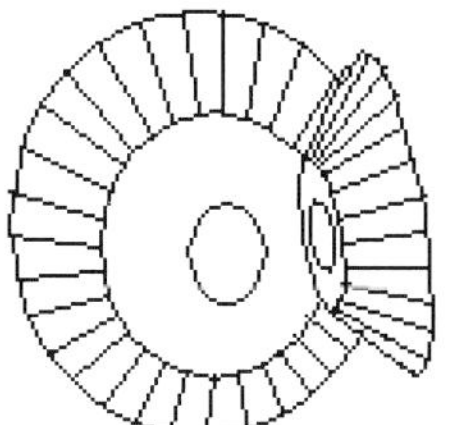

Neither parallel nor intersecting shafts

Crossed-helical gears

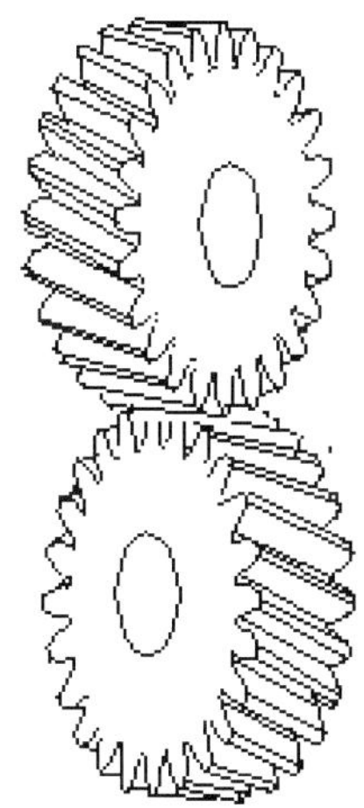

Worm and worm gear

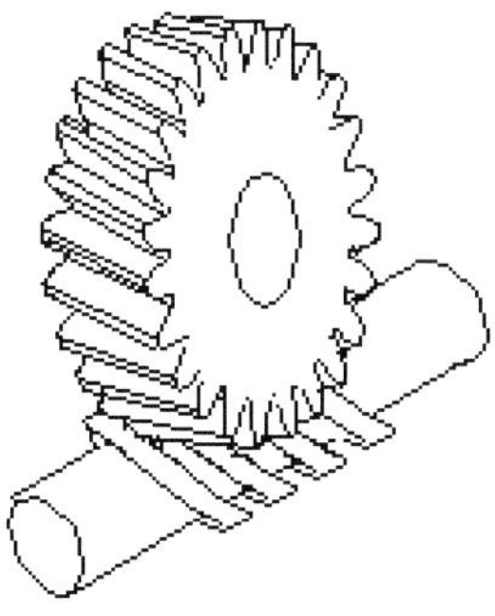

5. MOMENTUM, WORK, ENERGY, AND DYNAMICS

5.1 Momentum

Let's introduce some further concepts that will prove useful in describing motion. The first of these, *momentum*, was actually introduced by the French scientist and philosopher Descartes before Newton's time. Descartes' idea is best understood by considering a simple example: think first about someone weighing 45 kg standing motionless on high quality (frictionless) roller skates on a smooth, level floor. Someone standing in front of her, only a short distance away, throws a 5-kg medicine ball directly at her, so that we can assume that the ball's flight is essentially horizontal. She catches and holds the ball, and because of its impact, begins to roll backwards on the roller skates. Notice we've chosen her weight so that, conveniently, she and the ball together weigh just 10 times what the ball weighs by itself.

What is found in doing this experiment carefully is that after the catch, she and the ball roll backwards at just one-tenth the speed the ball was moving just before she caught it. So if the ball were thrown at 5 m/s, she would roll backwards at 1/2 m/s after the catch. It is tempting to conclude that the "total amount of motion" is the same before and after she catches the ball, since we end up with 10 times the mass moving at one-tenth the speed.

Considerations and experiments like this led Descartes to invent the concept of momentum, meaning "amount of motion," and to state that for a moving body, the momentum was the product of the mass of the body and its speed. Momentum is traditionally labeled by the letter p, so his definition, for a body having mass m and moving at speed v, was:

$$\text{momentum} = p = mv$$

It is obvious that in the above scenario of the woman catching the medicine ball, total momentum is the same before and after the catch. Initially, only the ball had momentum, an amount 5 x 5 = 25 in suitable units, since its mass is 5kg and its speed is 5 m/s. After the catch, there is a total mass of 50kg moving at a speed of 0.5 m/s, so the final momentum is 0.5 x 50 = 25. The total final amount is equal to the total initial amount. We have only randomly selected these figures, of course, but they reflect what is observed experimentally.

There is however a problem here. Obviously, one can imagine collisions in which the total amount of motion, as defined above, definitely is *not* the same before and after a collision. What about two people on roller skates, of equal weight, coming directly towards each other at equal but opposite velocities, who each come to a complete stop as they collide? Clearly in this situation there would be plenty of motion before the collision and none afterwards, so the total amount of motion definitely would not stay the same. In physics language, the total amount of motion is not conserved. Descartes was hung up on this problem a long time, but was rescued by a Dutchman, Christian Huygens, who pointed out that the problem could be solved in a consistent fashion if one did not insist that the quantity of motion be positive.

In other words, *if something moving to the right was taken to have positive momentum, then one should consider something moving to the left to have negative momentum.* With this convention, two people of equal mass coming together from opposite directions at the same speed would have total momentum *zero*, so if they came to a complete halt after meeting, as described above, the total momentum before the collision would be the same as the total after—that is, zero—and momentum *would* be conserved.

Of course, in the discussion above we are restricting ourselves to motions along a single line. It should be apparent that to get a definition of momentum that is conserved in collisions, what Huygens really did was to tell Descartes he should replace speed by *velocity* in his definition of momentum. It is a natural extension of this notion to think of momentum defined generally as:

momentum = mass x *velocity*

So, *since velocity is a vector, momentum is also a vector*, pointing in the same direction as the velocity, of course.

It turns out experimentally that in *any* collision between two objects (where no interaction with third objects, such as surfaces, interferes), the total momentum before the collision is the same as the total momentum after the collision. It doesn't matter whether the two objects stick together upon colliding or whether they bounce off each other, or what kind of forces they exert on each other, so conservation of momentum is a very general rule, quite independent of the details of the collision.

5.2 Momentum Conservation and Newton's Laws

As we have discussed above, Descartes introduced the concept of momentum and the general principle of conservation of momentum in collisions before Newton's time. However, it turns out that conservation of momentum can be deduced from Newton's laws. Newton's laws in principle fully describe all collision-type phenomena, and therefore must contain momentum conservation.

To understand how this comes about, consider first Newton's Second Law relating the acceleration a of a body of mass m with an external force F acting on it:

$$\text{force} = \text{mass x acceleration, or: } F = ma$$

Recall that acceleration is the rate of change of velocity, so we can rewrite the Second Law as:

force = mass x rate of change of velocity.

Now, the momentum is mv, mass times velocity. This means for an object having constant mass (which is almost always the case, of course),

the rate of change of momentum = mass x rate of change of velocity.

This means that Newton's Second Law can be rewritten:

force = rate of change of momentum.

Now think of a collision, or any kind of interaction, between two objects A and B. From Newton's Third Law, the force A feels from B is of equal magnitude to the force B feels from A, but in the opposite direction. Since (as we have just shown) force equals the rate of change of momentum, it follows that throughout the interaction process the rate of change of momentum of A is exactly opposite to the rate of change of momentum of B. In other words, since these are vectors, they are of equal length but pointing in opposite directions. This means that for every bit of momentum A gains, B gains the negative of that. In other words, B *loses* momentum at exactly the rate that A *gains* momentum, so their *total* momentum remains the same. But this is true throughout the interaction process, from beginning to end. Therefore, the total momentum at the end must be what it was at the beginning.

You may be thinking at this point: so what? We already know that Newton's laws are universally obeyed, so why dwell on one special consequence of them? The answer is that although we know Newton's laws are obeyed, this may not be much use to us in an actual case of two complicated objects colliding, because we may not be able to figure out what the forces are. Nevertheless, we *do* know that momentum will be conserved anyway. So if, for example, the two objects stick together, and no bits fly off, we can find their final velocity just from momentum conservation, without knowing any details of the collision.

5.3 Work

The word *work* as used in physics and engineering has a narrower meaning than it does in everyday life. First, it only refers to physical work, of course, and second, something has to be accomplished. If you lift up a box of books from the floor and put it on a shelf, you've done work, as defined in physics. If the box is too heavy and you tug at it until you're worn out but it doesn't move, that doesn't count as work.

Technically, work is done when a force pushes something and the object moves some distance in the direction it's being pushed. Consider lifting the box of books to a high shelf. If you lift the box at a steady speed, the force you are exerting is just overcoming gravity (experienced as the weight of the box). Otherwise the box would be accelerating. (Of course, initially you'd have to exert a little bit more force to get it going, and then at the end a little less as the box comes to rest at the height of the shelf.) It's obvious that you will have to do twice as much work to raise a box of twice the weight, so the work done is proportional to the force you exert. It's also clear that the work done depends on how high the shelf is. Putting these together, the definition of work is:

$$\text{work} = \text{force x distance}$$

where only distance traveled in the direction the force is pushing counts. With this definition, carrying the box of books across the room from one shelf to another of equal height doesn't count as work, because even though your arms have to exert a force upwards to keep the box from falling to the floor, you do not move the box in the direction of that force—that is, upwards. This may be a bit confusing, but it's important to grasp and to remember.

To get a more quantitative idea of how much work is being done, we need to have some units to measure work. Defining work as force times distance, we will measure distance in meters, but we've not yet talked about units for force. The simplest way to think of a unit of force is in terms of Newton's Second Law, force equals mass times acceleration.

The natural "unit force" would be that force which, pushing a unit mass (one kilogram) with no friction or other forces present, accelerates the mass at one meter per second per second (1 m/s^2), so that after two seconds the mass is moving at two meters per second, etc. This unit of force is called one *newton*. Note that a one-kilogram mass, when dropped, accelerates downwards at 10 meters per second per second. (Technically, the acceleration of gravity is 9.8 m/s^2, but we'll round off to 10 m/s^2 here to keep things simple.) This means that its weight, its gravitational attraction towards the earth, must be equal to 10 newtons. From this we can figure out that a one-newton force equals the weight of 100 grams—just less than a quarter of a pound, such as a stick of butter.

The downward acceleration of a freely falling object, 10 meters per second per second, is often written g for short (where g represents gravity). If we have a mass of m kilograms, we know it will accelerate it at g if it's dropped, so its weight is a force of magnitude mg, from Newton's Second Law.

Now back to *work*. Since work is force times distance, the natural unit of work would be the work done by a force of one newton over a distance of one meter. This unit of work is called a *joule*.

Finally, it is useful to have a unit for *rate of working*, also called *power*. The natural unit of rate of working is one joule per second, and this is called one *watt*. To get some feeling for rate of work, consider walking up stairs. A typical step is eight inches, or one-fifth of a meter, so you will gain altitude at, say, two-fifths of a meter per second. Assume your weight is 70 kg. Multiply that by 10 to get it in newtons; so it's 700 newtons. The rate of working then is 700 x 2/5, or 280 watts. Most people can't work at that rate for very long. A common English unit of power is the *horsepower*, which is 746 watts.

5.4 Energy and dynamics

Energy is the ability to do work.

5.4.1 Definitions and Units

In everyday conversation the words *force, energy,* and *power* are often used with similar meanings. But in science, engineering, and technology they are distinct numerical quantities, each with its own units.

Force is the quantity that causes, or tends to cause, a change in position or state of a body. If a body remains at rest (or moves at a constant velocity), the forces acting on it are in equilibrium and have no resultant effect. Force is a vector quantity and is measured in *Newtons* (N).

Suppose a force F (figure 9.1) causes a body to move a distance x in the direction of action. The force is said to do *work*. Work is a means of transferring energy by the movement of the point of application of a force through a distance in the direction of the force. Provided it is constant, the amount of work done or energy transferred is equal to the force multiplied by the distance moved, Fd. Work is a scalar quantity, not a vector quantity, since it has a magnitude but not a specified direction.

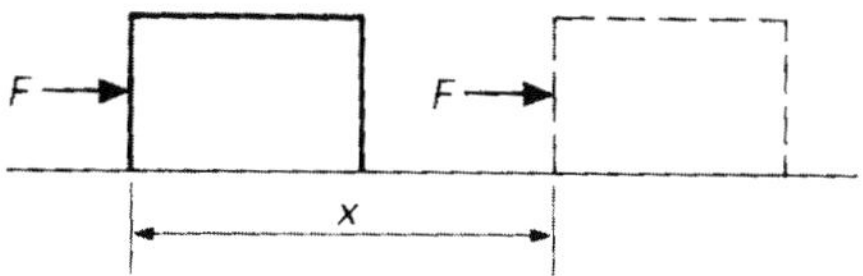

Figure 5.1

From the definition of work, its unit could be written newton-meter (Nm), but this is given the special name *joule,* and is denoted by J. (Nm is reserved for the unit of moment.) The joule is a very small unit of work, so values are often given in kilojoules (kJ), megajoules (MJ) or gigajoules (GJ).

We frequently use machines to help us do work. For personal transport we can use a bike or a gasoline-powered vehicle. We can cut mow the lawn either by pushing a hand mower or by using a gasoline-powered lawn mower. In these examples, the engines or motors do similar amounts of work to what we would do ourselves, but they can do it more quickly and with less effort on our part. We therefore judge engines and motors not by the total amount of work done but by the *rate* at which they do work. This is called *power,* and in SI units it is presented as the number of joules per second. This unit is given the name *watt* and is denoted by W. The units kilowatt (kW) and megawatt (MW) are more often used in practice for values of power. Power, like work, is a scalar quantity possessing magnitude but not direction.

Suppose the force F in figure 5.1 moves thought the distance d in time t at a constant speed v, then by definition, the power P is:

$$P = \text{rate of doing work} = \frac{\text{work done}}{\text{time taken}} = \frac{\text{Fx}}{\text{t}} = Fv$$

since speed is equal to distance divided by time. If the speed is changing, the last expression gives the instantaneous value of the power.

The unit *horsepower*, introduced by James Watt, is widely used. It is abbreviated hp, and 1 hp = 746 W.

5.4.2 Types of Energy

In order for work to be done by machines or human beings, they need to possess energy. Energy is measured in the same units as work, namely the joule (J), with values often being expressed, for practical purposes, as kilojoules (kJ), megajoules (MJ) or gigajoules (GJ). Energy, like work, does not possess direction, and is a scalar quantity. It can take several forms and may be converted from one form to another.

Kinetic Energy

This is the energy that a body posses because of its motion. The greater the speed at which it is moving, the greater its kinetic energy. Linear motion and angular motion give rise to kinetic energy.

Potential Energy

This is the energy that a body possesses because of its position in a particular type of field: gravitational, electric, magnetic, or nuclear. Gravitational potential energy, for example, equals the amount of work that a body can do as it falls towards the earth under the pull of gravity. Potential energy, like kinetic, can be used to do useful work. In a hydroelectric generating station, for instance, water falls from a higher level to a lower one and does work on the turbines it passes through.

Kinetic and Potential Energy in Depth

It takes work to drive a nail into a piece of wood. A force has to push the nail a certain distance, against the resistance of the wood. A moving hammer, hitting the nail, can drive it in. A stationary hammer placed on the nail does nothing. The moving hammer has energy—the ability to drive the nail in—because it's moving. This hammer energy is called kinetic energy. Kinetic is just the Greek word for motion, and is the root word for cinema, meaning *movies*.

Another way to drive the nail in, if you have a good aim, might be to simply drop the hammer onto the nail from some suitable height. By the time the hammer reaches the nail, it will have kinetic energy. It has this energy, of course, because the force of gravity (its weight) accelerated it as it came down. But this energy didn't come from nowhere. Work had to be done in the first place to lift the hammer to the height from which it was dropped onto the nail. In fact, the work done in the initial lifting, force x distance, is just the weight of the hammer multiplied by the distance it is raised, in joules. But this is exactly the same amount of work as gravity does on the hammer in speeding it up during its fall onto the nail. Therefore, while the hammer is at the top, waiting to be dropped, it can be thought of as storing the work that was done in lifting it, which is ready to be released at any time. This "stored work" is called *potential energy*, since it has the *potential* of being transformed into kinetic energy just by releasing the hammer.

To give an example, suppose we have a hammer of mass 2 kg, and we lift it up through 5 meters. The hammer's weight, the force of gravity, is 20 newtons (recall it would accelerate at 10 meters per second per second under gravity, like anything else), so the work done in lifting it is force x distance = 20 x 5 = 100 joules, since lifting it at a steady rate requires a lifting force that just balances the weight. This 100 joules is now stored ready for use—that is, it is potential energy. Upon releasing the hammer, the potential energy becomes kinetic energy. The force of gravity pulls the hammer downwards through the same distance the hammer was originally raised upwards, so since it's a force of the same amount as the original lifting force. The work done on the hammer by gravity in giving it motion is the same as the work done previously in lifting it, so as it hits the nail it has a kinetic energy of 100 joules. We say that the potential energy is transformed into kinetic energy, which is then spent driving in the nail.

We should emphasize that both energy and work are measured in the same units, joules. From the above discussion, a mass of m kilograms has a weight of mg newtons. It follows that the work needed to raise it through a height h meters is force x distance, that is weight x height, or mgh joules. This is the potential energy.

Historically, this was the way energy was stored to drive clocks. Large weights were raised once a week, and as they gradually fell, they released energy, turned the gears, and by a sequence of ingenious devices, kept the pendulum swinging. The problem was that this necessitated rather large clocks to get a sufficient vertical drop to store enough energy, so spring-driven clocks became more popular once they were developed. Compressing a spring is just another way of storing energy. It takes work to compress a spring, but (apart from small frictional effects) all that work is released as the spring uncoils or springs back. The stored energy in the compressed spring is often called *elastic potential energy*, as opposed to the *gravitational potential energy* of the raised weight.

Above, we've given an explicit way to find the potential energy increase of a mass m when it's lifted through a height h. It's just the work done by the force that raised it, force x distance = weight x height = mgh.

Kinetic energy is created when a force does work accelerating a mass and increases its speed. Just as for potential energy, we can find the kinetic energy created by figuring out how much work the force does in speeding up the body.

Remember that a force only does work if the body the force is acting on moves in the direction of the force. For example, for a satellite going in a circular orbit around the earth, the force of gravity is constantly accelerating the body downwards, but it never gets any closer to sea level, it just swings around. Thus, the body does not actually move any distance in the direction gravity's pulling it, and in this case gravity does no work on the body.

Consider, in contrast, the work the force of gravity does on a stone that is simply dropped from a cliff. Let's be specific and suppose it's a one-kilogram stone, so the force of gravity is 10 newtons downwards. In one second, the stone will be moving at 10 meters per second, and will have dropped five meters. The work done at this point by gravity is force x distance = 10 newtons x 5 meters = 50 joules, so this is the kinetic energy of a one-kilogram mass going at 10 meters per second.

How does the kinetic energy increase with speed? Think about the situation after 2 seconds. The mass has now increased in speed to 20 meters per second. It has fallen a total distance of 20 meters (average speed 10 meters per second x time elapsed of 2 seconds). So the work done by the force of gravity in accelerating the mass over the first two seconds is force x distance = 10 newtons x 20 meters = 200 joules.

So we find that the kinetic energy of a one-kilogram mass moving at 10 meters per second is 50 joules, moving at 20 meters per second it's 200 joules. It's not difficult to check that after three seconds, when the mass is moving at 30 meters per second, the kinetic energy is 450 joules. The essential point is that the speed increases linearly with time, but the work done by the constant gravitational force depends on how far the stone has dropped, and that goes as the square of the time. Therefore, the kinetic energy of the falling stone depends on the square of the time, and that's the same as depending on the square of the velocity. For stones of different masses, the kinetic energy at the same speed will be proportional to the mass (since weight is proportional to mass, and the work done by gravity is proportional to the weight), so using the figures we worked out above for a one-kilogram mass, we can conclude that for a mass of *m* kilograms moving at a speed *v* the kinetic energy must be:

kinetic energy = ***1/2mv_***

Strain Energy

When a force is applied to a structure and causes it to distort, the point of application of the force is changed and work is done. This can be stored in the structure as strain energy or resilience. If the force is removed, this energy may be recovered and converted into other forms. In the operation of a mechanical pinball machine, for instance, strain energy is stored in the spring as the plunger is pulled back, and then released in the form of kinetic energy or potential energy of the ball.

Heat

All substances consist of molecules that are in motion, and between which there are forces of attraction. We can think of the molecules as possessing kinetic energy and potential energy corresponding to the two forms described above. In everyday language we call this molecular energy heat. In thermodynamics it is termed *internal energy* and the word "heat" is used to mean the energy that is transferred from one body to another by conduction, convection or radiation.

Chemical Energy

Fuels such as oil and coal contain the elements carbon, hydrogen, and sometimes sulfur, each of which combines with oxygen when the fuel is burned, releasing energy in the process. The amount of energy available from the fuel is called its *calorific value.* The energy released during combustion may be used directly for heating purposes or it may be converted to other forms.

Electrical Energy

Electricity is not a form of energy itself but it is often a very convenient means of transferring energy from one body to another. We use the term "electrical energy" for the energy being transferred. The energy transferred in this way can be stored in electrochemical batteries as potential energy.

5.4.3 Units of Energy

The recommended SI unit, the joule (J), should be used in all technical calculations. However, for historical reasons, several other units are still used, and it is often necessary to convert values to joules.

Kilowatt-hour

The electricity industry uses this unit for measuring the energy supplied to its customers. It is abbreviated to kWh and represents the energy that corresponds to a power of 1 kW operating for one hour. Since 1 h= 3600 s and 1 W=1 J/s then 1 kWh = 1000 X 3600 J = 3.6 MJ.

British Thermal Unit (BTU) and the Therm

These are heat units based on the pound and the Fahrenheit scale of temperature. The British thermal unit (abbreviated BTU) was defined as the heat required to raise the temperature of 1 lb of water by 1 degree Fahrenheit. For most purposes it is sufficiently accurate to consider that 1 BTU = 1055 J = 1.055 kJ.

The gas industry previously used the therm as a unit of heat. This is defined as 100,000 BTU, and therefore 1 therm = 10^5 X 1055 J = 105.5 X 10^6 J = 105.5 MJ. The industry now uses the kWh. One therm = 29.3 kWh. This enables a direct comparison to be made between gas and electricity, but it must be remembered that heating appliances are not 100 percent efficient.

Calorie

The relationship between heat and energy was not fully understood until the middle of the nineteenth century. By this time the unit calorie had been widely adopted by scientists for quantities of heat. It is abbreviated *cal,* and was originally defined as the heat required to raise 1 g of water by 1 degree Celsius, but this is not a precise definition. By international agreement in 1956 the value of the calorie was fixed as 4.1868 J. Nowadays it is used mainly in the form kilocalorie (kcal) for the energy values of food. In summary:

1 kWh = 3.6 MJ
1 BTU = 1.055 kJ
1 therm = 105.5 MJ
1 cal = 4.1868 J

5.4.4 Sources of Energy

Solar Energy

Enormous amounts of energy from the sun reach the outer atmosphere of the earth. The quantity reaching the surface depends on the ozone layer, dust, clouds, and latitude. Even in temperate latitudes it can amount to 500 W/m^2. This energy is readily available for producing hot water and space heating using simple solar panels. Installation costs are high, however, and the energy available decreases in winter when the requirement is greatest. Much higher temperatures are needed for power generation, and these can't be achieved simply by focusing the sun's rays with concave mirrors.

Solar cells convert the sun's energy directly into electricity. They have been used to power small experimental cars through electric motors of about 1hp, and roadside warning signs. Photoelectric cells are already widely used in light meters and solar powered calculators and to recharge batteries where very small currents are sufficient, but they are not yet economic for large power generation.

Ocean Thermal Energy

The sun causes the surface temperatures of oceans to be considerably higher than at great depths. This temperature difference could be used in a thermodynamic cycle to generate electricity.

Tidal Energy

Seawater can be used to drive turbines in a barrage built across a bay or estuary using the potential energy that exists when the levels are difference on the two sides. In its simplest form, such a scheme could only generate electricity for part of the day, but constant generation is possible using two basins. One tidal generating scheme has been built over a river estuary in northern France.

The use of tidal energy is not new. The tidal mill at Woodbridge in Suffolk had been working for 200 years when it closed in 1957, and the first such mill on the site was built in about 1170 AD.

Hydroelectric Power

In mountainous countries substantial amounts of power can be generated from the potential energy of water stored by dams. Although the construction costs can be very high, the direct cost of production is very small.

Wave Energy

The vertical motion of the sea can be utilized to generate power in several ways. One system uses the relative movements of floating rafts to convert the wave energy. Very large amounts of energy are theoretically available, particularly in the winter when the requirement is greatest.

Geothermal Energy

We know that very high temperatures exist within the Earth, and in some places heat is even available in sufficient at the Earth's surface. Hot springs have been used at least since Roman times. There are installations in several countries that use geothermal energy to generate electricity.

Wind Energy

Wind power has been harnessed for many centuries and is now seriously considered for the generation of electricity. The largest windmills (or aerogenerators) in use at the present time have a diameter of about 50 m and an output of roughly 1 MW. Harnessing the energy of the wind generates no pollution as a byproduct, and consumes no non-renewable resources.

Biomass Energy

Plants can yield energy in several ways and, unlike fossil fuels, are renewable. Wood, for example, still is an important fuel in many countries and, under favorable conditions, the tress on about one hectare (10,000 m^2) can produce sufficient timber to meet the energy requirements of an average house on a continuous basis.

Instead of burning plants to obtain energy, we also can obtain fuels such as methane, ethanol, and methanol as energy sources from them.

Although these alternative sources of energy can reduce our dependence on fossil fuels, they have their own limitations and disadvantages. Most of them involve large construction costs, which may outweigh the savings in fuel costs. Furthermore, they are not all suitable for every country. Solar energy, for example, requires a favorable climate, and hydroelectric power is only available in significant amounts in mountainous regions of the world. Winds and tides vary from season to season and day to day, so their energy is not always available when it is needed.

Exercises:

5.1 Both momentum and kinetic energy are in some sense measures of the amount of motion of a body. How do they differ?

5.2 Can a body change in momentum without changing in kinetic energy?

5.3 Can a body change in kinetic energy without changing in momentum?

5.4 Suppose two lumps of clay of equal mass traveling in opposite directions at the same speed collide head-on and stick to each other. Is momentum conserved? Is kinetic energy conserved?

5.5 As a stone drops off a cliff, both its potential energy and its kinetic energy continuously change. How are these changes related to each other?

6. FLUID MECHANICS, PNEUMATIC AND HYDRAULIC SYSTEMS

Fluid power uses pressurized fluids (liquids and gasses) to control and transmit power. These fluids travel through fluid lines from a reservoir to a location where the power is needed to do work. Fluids react to changes in temperature, force, and volume. *Hydraulics* is the term used to describe systems in which liquids are controlled. *Pneumatics* is the term used to describe when gasses are controlled.

Atmospheric pressure and vacuum also are important terms to discuss. Atmospheric pressure is described as:

> "Pressure caused by the weight of the atmosphere. At sea level it has a mean value of one atmosphere but reduces with increasing altitude". *American Heritage(r) Dictionary of the English Language, Fourth Edition, Copyright (c) 2000 by Houghton Mifflin Company*

While you are sitting at your desk reading this, there are roughly 14.7 pounds per cubic inch of air pressure on you. This figure varies with change in elevation from sea level. You may have a pressure gauge that helps you determine how much pressure is in your bike or car tire. These gauges are set to initially reflect 0 psi (pounds per square inch), meaning that it is equal to the atmospheric pressure present. In fluid power, pressure is measured when something is in motion. However, no motion can happen until fluid power overcomes atmospheric pressure.

Liquids have weight associated with them. If we were to look at water in a swimming pool, we could define the weight created by the depth (or height) of the water. The pressure created by a column of water is called static head pressure, or static pressure. The deeper you go in the pool, the higher the pressure will be. Imagine a column of water 28 inches high. This column will exert one pound per square inch of pressure (psi) down on the bottom. This column can be very wide or very narrow. It could be a 1" pipe, or a 3' pipe. If the water is 28 inches deep, there's going to be one pound per square inch of pressure at the bottom, no matter what diameter the column of water is.

SCUBA is the acronym for self-contained underwater breathing apparatus. SCUBA divers use gauges to determine how much pressure is on them at any given time. As they dive deeper, the pressure increases. As they come to the surface, the pressure decreases. There are very specific rules in SCUBA diving for how quickly divers may come back up to the surface to keep the levels of nitrogen in their bloodstream from getting too high.

10.1 Pneumatics

Gasses do not have a specific volume as liquids do. Because gasses have no specific volume, they expand to fill up the volume of the container that they are in. Gas molecules can, however, be forced together. This is called *compression.* Boyle's Law states that the volume of a gas varies inversely with the pressure applied to it, provided that the temperature remains constant. Charles' Law states that a volume of gas is directly proportional to the temperature. This means that a change in temperature, no matter how small, results in a change of volume of the gas, provided pressure stays the same.

Pneumatic systems require that air be compressed in order to do work. A compressor is used to create the high air pressure. The two most common types of compressors are reciprocating piston and sliding vane. A piston-type compressor draws air in from a source and compresses it into a smaller space. A sliding vane compressor is much like a historical riverboat's paddle wheel. It has paddles or "vanes" connected to a rotor that draws air in from a source and compresses it into a high-pressure air outlet. Both systems compress air into a high-pressure storage space known as a *receiver.*

All pneumatic systems need *regulators* to keep the pressure at a constant level. When the air is moved in from a source, moisture may accompany it. For this reason, a dryer is needed to remove moisture from the supply line.

10.2 Hydraulics

Liquids, as we looked at earlier, have a definite volume. These volumes cannot normally be compressed—a fact that allows liquids to transfer force efficiently. When transferring force though a liquid, a mechanical device (usually a piston) puts pressure on the fluid. We can harness this force by limiting the area that the liquid can act upon, as long as there is room for the liquid to flow. Changing the size of the container does not affect the ability of the liquid to transmit force. Input and output forces are always equal, if the fluid is free flowing.

The *hydraulic lift* shown in figure 10.1 is a lifting machine that uses fluid pressures. In this lift the amount of fluid displaced by the effort piston must be equal to the amount of fluid displaced at the load piston. So the volumes A_1x and A_2y must be the same.

$$\text{Therefore} \quad VR = \frac{x}{y} = \frac{A_2}{A_1}$$

The area of the load piston is substantially larger than that of the effort piston, so a small effort applied to a hydraulic jack will lift a large load.

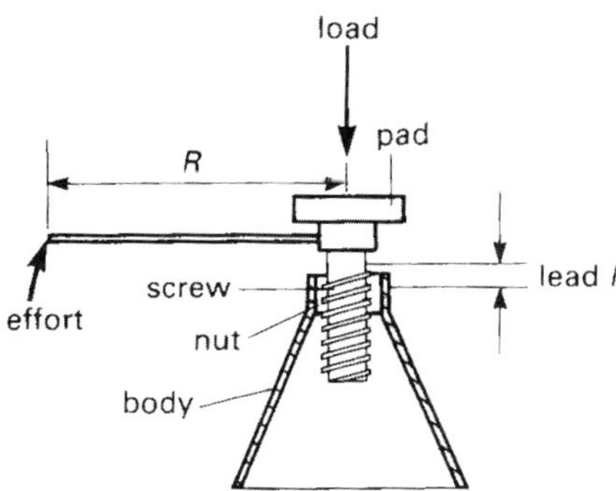

Figure 10.1 the hydraulic lift

7. THERMODYNAMICS

Thermodynamics is a branch of physics that deals with the energy and work of a system. the first developed in the 19th century as scientists were first discovering how to build and operate steam engines, thermodynamics deals only with the large-scale response of a system that we can observe and measure in experiments. Small-scale gas interactions are described by the kinetic theory of gases. The methods compliment each other; some principles are more easily understood in terms of thermodynamics and some principles are more easily explained by kinetic theory.

There originally were three principal laws of thermodynamics, but in the early 20th century, it became clear that a fourth, more fundamental law needed to be explicitly stated. So, since this more fundamental law came into use after the other three were already widely in use and named as such, Ralph H. Fowler coined the term "zeroth law" (as in zeroth, first, second, third) of thermodynamics to indicate that this newly identified law actually preceded the other three laws in terms of its theory.

Temperature

An important property of any liquid, solid, or gas is *temperature*. We have some experience with temperature that we don't have with properties like viscosity and compressibility. We've heard the TV meteorologist give the daily value of the temperature of the atmosphere (75 degrees, for example). We know that a hot object has a high temperature, and a cold object has a low temperature. And we know that the temperature of an object changes when we heat the object or cool it.

Scientists, however, must be more precise than simply describing an object as "hot" or "cold." An entire branch of physics, called thermodynamics, is devoted to studying the temperature of objects and the transfer of heat between objects of different temperatures.

References

(Berard *et al.*55) Berard, S. J., Waters, E. O. and Phelps, C. W., *Principles of Machine Design*, Ronald Press, New York, 1955.

(Bradford & Guillet 43) Bradford, L. J. and Guillet, G. L., *Kinematics and Machine Design*, John Wiley and Sons, New York, 1943.

(Chace) Chace, M. A., *ADAMS and DRAM, Automatic Dynamic Analysis of Mechanical Systems and Dynamic Response of Articulated Machinery, University of Michigan.*

(Chace 63) Chace, M. A., "Vector Analysis of Linkages," Journal of Engineering in Industry, Series B, vol. 55, no. 3, August 1963, pp.289-297.

(Chen 82) Chen, F. Y., Mechanics and Design of Cam Mechanisms, Pergamon Press, New York, 1982.

(Dean 62) Dean, P. M. Jr., "Gear-Tooth Proportions," Chapter 5 in Gear Handbook, ed. D. W.Dudley, McGraw-Hill, New York, 1962.

(Denavit & Hartenberg 55) Denavit, J. and Hartenberg, R. S., "A Kinematic Notation for Lower-Pair Mechanisms Based on Matrices," ASME Journal of Applied Mechanisms, 1955, pp. 215-221.

(Erdman & Sandor 84)Erdman, A. and Sandor, G., Mechanism Design: Analysis and Synthesis, Prentice-Hall, New Jersey, 1984.

(Finger et al. 92) Finger, S., Fox, M. S., Prinz, F. B. and Rinderle, J. R., "Concurrent Design," Applied Artificial Intelligence, Vol. 6, 1992, pp. 257-283.

(Hale 15) Hale, J. W. L., Practical Mechanics and Allied Subjects, McGraw-Hill Book Company, 1915.

(Hall 66) Hall, A. S., Jr., Kinematics and Linkage Design, Balt Publisher, West Lafayette, IN, 1966.

(Ham et al. 58) Ham, C. W., Crank E. J. and Rogers W. L., Mechanics of Machinery, McGraw-Hill, 1958.

(Hartenberg and Denavit 80) Hartenberg, R. S. and Denavit, J., Kinematic Synthesis of Linkages, McGraw-Hill, New York, 1980.

(Hunt 78) Hunt, K. H., Kinematic Geometry of Mechanisms, Oxford University Press, New York, 1978.

(Kepler 60) Kepler, H. B., Basic Graphical Kinematics, MaGraw-Hill, New York, 1960.

Norman, E., Cubitt, J., Urry, S., & Whittaker, M. (1995). *Advanced Design and Technology* (2nd ed.). England: Longman.

(Patton 79) Patton, W. J., Kinematics, Reston Publishing, 1979.

(Paul 79) Paul, B., Kinematics and Dynamics of Planar Machinery, Prentice-Hall, New Jersey, 1979.

(Sandor & Erdman 84) Sandor, G. and Erdman, A., Advanced Mechanism Design: Analysis and Synthesis, Prentice-Hall, New Jersey, 1984.

(Sheth & Uicker) Sheth, P. N. and Uicker, J. J., "IMP (Integrated Mechanisms Program), A Computer-Aided Design Analysis System for Mechanisms and Linkages," ASME Journal of Engineering for Industry, May, 1972, pp.454-464.

(Shigley & Uicker 80) Shigley, J. E. and Uicker, J. J., Jr., Theory of Machines and Mechanisms, McGraw-Hill, New York, 1980.

(Shigley et al. 86) Shigley, J. E., Mischke, C. R. et al., Standard Handbook of Machine Design, McGraw-Hill, New York, 1986.

(Shu 78) Shu, C. H., Kinematics and Mechanisms Design, John Wiley & Sons, 1978.

(Uicker et al. 64) Uicker, J. J., Jr., Denavit, J. and Hartenberg, R. S., "An Interactive Method for the Displacement Analysis of Spatial Mechanisms," ASME Journal of Applied Mechanics, Vol. 86, ser. E, 1964, pp. 309-314.

(Uri et al. 68) Uri, Haber-Schaim, et al., College Physics, Paytheon Education, 1968.

(Yang & Freudenstein 64) Yang, A. T. and Freudenstein, F., "Application of Dual-Number and Quaternion Algebra to the Analysis of Spatial Linkages," ASME Journal of Applied Mechanics, Vol. 31, Series E, 1964, pp. 300-308.

Section References

Sections 2.2.1, 2.3.1, 2.3.2, 2.3.3, 2.4, 2.5, 4.1, 9.4.1- 9.4.4
were contributed courtesy of the Longman Publishing Group of Essex England.
Norman, E., Cubitt, J., Urry, S., & Whittaker, M. (1995). *Advanced Design and Technology* (2nd ed.). England: Longman.
Sections , 1, 2.1, 2.2, 2.6, 2.7, 3, 4.1.1, 4.2.2. 4.2-4.6, 5, 6, 7, 8
were contributed and reprinted by permission of Yi Zhang, Susan Finger and Stephannie Behrens
Rapid Design through Virtual and Physical Prototyping
Carnegie Mellon University.
Copyright Material, reproduced by permission of the author http://www.cs.cmu.edu/People/rapidproto/mechanisms
Sections 9.1-9.3 and the Kinetic and potential energy in depth sections were contributed by Michael Fowler, University of Virginia Physics. Visit his website for more information and additional lectures:
http://galileoandeinstein.physics.virginia.edu/lectures/momentum.html
Materials in section 11 were contributed by NASA's Glenn Research center. Visit NASA's Glenn Research Center's website for more information on Thermodynamics and for interactive lessons, graphics, details and simulators on Thermodynamics. http://www.grc.nasa.gov/WWW/K-12/airplane/thermo.html

Glossary

ABET Accreditation Board for Engineering and Technology.

Abstract A statement summarizing the important points of a given text.

Acceleration The variation of velocity over a specified period of time.

Acoustical Engineer Plans, perfects, or improves the sound of an architectural space.

Aero Engineer Responsible for developing planes, missiles, and spacecraft.

Agricultural Engineer Designs and develops the means by which food is produced free from contamination.

Alternating Current (AC) Electric current that reverses direction, usually many times per second. Most electrical generators produce alternating current.

Ampere (Amp) A measure of electrical current flow.

Analysis Engineer Function performed in conjunction with design, development, and research using mathematical models and computational tools to provide information for design, development, or research engineers.

Applet A program that is typically stored on a remote computer that users can connect to through the Web.

Automotive Engineer Plans, coordinates, and implements the specifications for a new car's parts.

Bio Engineer Applies engineering principles to biological systems.

Brainstorming A technique used to stimulate as many innovative solutions as possible.

Browser Software that allows users to access and navigate the Internet.

C++ (C plus plus) A popular object-oriented programming language developed in 1985.

Cam A cam is a projecting part of a rotating wheel or shaft that strikes a lever at one or more points on its circular path.

Capacitor An electronic device which stores energy and releases it when needed.

Celsius Scale Temperature scale using the freezing point of water as the zero point and the boiling point of water as 100 degrees.

Ceramics Family of materials made by shaping and then firing a nonmetallic mineral at high temperature.

Chemical Engineer Applies fundamental engineering, chemistry, and physics principles to design and develop processes to produce products.

Civil Engineer Responsible for designing and supervising the construction of roads, buildings, airports, tunnels, bridges, and water and sewage systems.

Composite A combination of two or more constituent materials bonded together in an effort to produce a material with superior properties to either of the individual materials.

Computer Engineer Designs and builds computer-related hardware products for a variety of applications.

Construction Engineer Concerned with the management and operations of construction projects.

Construction Engineer Turns the construction specifications of a Design Engineer into reality.

Consulting Engineer Provides technical expertise to organizations that provide goods and services.

Current The flow of electric charge.

Design (See Engineering Design)

Design Engineer Responsible for the design of a particular component or part of a product and the relevant detailed specifications.

Development Engineer Takes knowledge acquired by Research Engineer and applies it to a specific product or application.

Direct Current (DC) An electrical current which flows only in one direction in a circuit. Batteries and fuel cells produce direct current.

Displacement The distance between the initial and final position of an object.

Efficiency The ratio of the output to the input in any system.

Electrical Engineer Responsible for the design, development, testing, and supervision of the manufacture of electrical equipment. More numerous than engineers of any other discipline.

Energy The capacity of a physical system to do work.

Engineering The profession in which the knowledge of math and science is applied to develop ways to utilize materials and the forces of nature for the benefit of people.

Engineering Design The process of devising a system, component, or process to meet desired needs.

Engineering Technology The part of the technological field which requires the application of scientific and engineering knowledge and methods combined with technical skills in support of engineering activities.

Environmental Engineer Applies engineering principles to improve and maintain the environment.

Ethics Standards, rules, or guidelines for morally or socially approved conduct.

Executive Summary Condenses a larger work into essentials for efficient use from a managerial perspective.

Fahrenheit Scale Temperature scale based on 32F for the temperature at which water freezes and 212F for the temperature at which water boils (180 difference).

File Transfer Protocol (ftp) Program that allows users to transfer files on a network.

Fire Protection Engineer Designs fire sprinkler and alarm systems as well as exit and control systems.

Food Processing Engineer Concerned with providing healthier products for consumers.

Force An agent of influence that, if applied to an object, results mainly in an acceleration of the object.

Fulcrum A pivot about which a lever turns.

Fulcrum The pivot about which a lever turns.

Gantt Chart Project managerial tool used to plan the work requirements and schedules and direct the use of resources.

Gear A gear is a toothed wheel that engages another toothed mechanism in order to change the speed or direction of transmitted motion.

Genetic Engineer Uses scientific methods to research genes found in the cells of plants and animals to develop better products, medicines, and services.

Geological Engineer Uses physics to map and predict the flow of water in rivers and above and underground. Takes measures to help prevent landslides and build structures in rock. Builds and maintains earth-related power sources.

Heat A form of energy that is transferred by a difference in temperature.

Hydraulics Refers to a system in which liquids are controlled.

Industrial Engineer Designs, improves, and installs integrated systems of people, material and energy. Traditionally done on a factory floor.

Inertia The tendency of an object at rest to remain at rest, and of an object in motion to remain in motion, outlined in Newton's First Law of Motion.

Input The instructions or data used by a system to carry out a task.

Internet A worldwide network of computers that allows electronic communication by sending packets of data back and forth throughout the network.

IP Address A unique 32-bit address assigned to each organization that becomes part of the Internet.

Lever A rigid bar pivoted about a fulcrum (pivot).

Machine A mechanical or electrical device that transmits or modifies energy to perform or assist in the performance of useful work.

Manufacturing Engineer Applies and controls engineering procedures in manufacturing processes.

Marine and Ocean Engineer Concerned with the exploration of the oceans, the transportation of products over water, and the utilization of resources in the world's oceans, seas, and lakes.

Materials Science Engineer Develops new materials and the processes to create them.

Mechanical Advantage The factor by which a machine multiplies the force put into it.

Mechanical Engineer Involved in the design, development, production, control, operation, and service of machines and mechanical devices. A member of one of the largest and broadest of all engineering disciplines. Outnumbered only by Electrical Engineers.

Mechanism A system that transforms one kind of motion to another, and may consist of a single component or a combination of the fundamental components: lever, cam, screw pulley, gear, or ratchet.

Memo Convenient, relatively informal way to communicate the existence of a problem, propose some course of action, or report the results of a test.

Metals Elements that typically have a shiny surface, are generally good conductors of heat and electricity, and can be melted or fused, hammered into thin sheets, or drawn into wires.

Mineral and Mining Engineer Responsible for maintaining the flow of raw materials by discovering, extracting, and processing minerals.

Momentum The mass of an object multiplied by its velocity. If an object has a high momentum, it is more difficult to change its velocity.

Moore's Law The capacity of computer chips doubles approximately every two years.

Motion A change of position or orientation.

Myers Briggs Personality assessment questionnaire.

Newton's First Law of Motion (Inertia) The tendency of an object at rest to remain at rest, and of an object in motion to remain in motion.

Newton's Second Law of Motion The acceleration of an object is directly proportional to the force exerted on the object, and is inversely proportional to the mass of the object (in other words, as mass increases, the acceleration has to decrease). The acceleration of an object moves in the same direction as the total force.

Nuclear Engineer Studies nuclear energy and radiation and their beneficial uses.

Ohm A measure of how much something resists (impedes) the flow of electricity. Larger numbers mean more resistance.

Ohm's Law A statement of the relationship between the voltage across an electrical component and the current through the component written as $V=IR$.

Operating System A group of programs that manage the operation of a computer.

Operations Engineer Oversees the ongoing facility performance.

Optimization Doing the most with the least.

Output The final product of a system or device.

Petroleum Engineer maintains the flow of petroleum in a safe and reliable manner.

Pneumatics Refers to a system in which gases are controlled.

Potential Energy The capacity to do work due to the position of an object.

Power The rate at which work is done.

PowerPoint A popular electronic presentation software package.

Process The operations in a system that produces something.

Proposal Describes a need or problem at issue, defines a solution, and requests funding or other resources to solve the problem.

Protocols Used to move data from one location to another across the Internet while verifying that the data transfer was successful and complete.

Request For Proposal (RFP) Specification of requirements pertaining to a project which is sent to suppliers who reply with proposals.

Research Engineer Explores the fundamental principles of chemistry, physics, biology, and mathematics to overcome barriers to advancement in their field.

Resistor An electronic component that opposes the flow of electrical current, measured in ohms.

Robotics Engineer Designs efficient and skilled robots in order to assemble complex products and parts.

Rube Goldberg Machine A project which incorporates outlandish tools and steps to accomplish simple everyday tasks in a most complex, innovative way.

Scientific Method A basis for solving a problem in which one defines the problem, gathers facts, develops a hypothesis, performs a test, and evaluates the results.

Software Engineer Solves problems in a program development.

Spreadsheet A screen-oriented interactive program enabling a user to lay out financial data on the screen.

Structural Engineer Creates safer, more efficient structures.

System A means of achieving a desired result. Involves input, process, output, and feedback.

Systems Engineer Works with overall design, development, manufacture, and operation of a complete system or product.

Technical Support Engineer Serves as the link between customer and product, assisting with installation and setup.

Temperature The degree of hotness or coldness of a body or environment (corresponding to its molecular activity).

Test Engineer Responsible for designing and implementing tests to verify the integrity, reliability, and quality of products.

Torque Something that produces or tends to produce rotation, and whose effectiveness is measured by multiplying the force times the perpendicular distance from the line of action of the force to the axis of rotation.

Velocity A measurement of the rate and direction of motion.

Volt A basic unit of electrical potential. One volt is the force required to send one ampere of electrical current through a resistance of one ohm.

Voltage A measure of electrical potential measured in volts.

Work A force applied over a distance.

Index

E

F

G

H

I

J

L

M

N

O

P